FOOD AND BEVERAGE CONSUMPTION AND HEALTH

PROANTHOCYANIDINS

FOOD SOURCES, ANTIOXIDANT PROPERTIES AND HEALTH BENEFITS

FOOD AND BEVERAGE CONSUMPTION AND HEALTH

Additional books in this series can be found on Nova's website under the Series tab.

Additional e-books in this series can be found on Nova's website under the e-book tab.

FOOD AND BEVERAGE CONSUMPTION AND HEALTH

PROANTHOCYANIDINS

FOOD SOURCES, ANTIOXIDANT PROPERTIES AND HEALTH BENEFITS

INGRID SULLIVAN
EDITOR

nova publishers
New York

Library of Congress Cataloging-in-Publication Data

ISBN: 978-1-63463-430-4
Library of Congress Control Number: 2014954516

Published by Nova Science Publishers, Inc. † New York

CONTENTS

PREFACE

Proanthocyanidins (PACs), present in a wide variety of food products and beverages, are the second most abundant class of phenolic compounds in our diet. These polyphenolic compounds are found in high proportions in cocoa products, seeds and skins of grapes, green and black tea, cinnamon and berries, among others. These compounds possess multiple therapeutic effects including anticancer, antimicrobial activity, cardioprotective role, prevention of atherosclerosis, and neuroprotective action. Many of these pharmacological benefits, including their neuroprotective action, are attributed to their significant antioxidant activity as free radical scavengers and as modulators of oxidant/antioxidant status. This book discusses food sources, the antioxidant properties and health benefits of proanthocyanidins.

Chapter 1 - Proanthocyanidins (PACs), present in a wide variety of food products and beverages, are the second most abundant class of phenolic compounds in our diet. The most common PACs are procyanidins, exclusively consisting of (epi)catechin units. Besides procyanidins, prodelphinidins, a heterogeneous group consisting of at least one (epi)gallocatechin unit and additional (epi)catechin units, and propelargonidins, a heterogeneous group consisting of at least one (epi)afzelechin unit and additional (epi)catechin units, occur in food sources. PACs consist of monomeric flavan-3-ol units, which are mainly linked through C4-C8 or sometimes C4-C6 bonds. These linkages are both called B-type linkages. When an additional ether linkage is formed (mainly) between C2-O-C7, compounds are called A-type PACs.

The most well-known sources of A-type PACs are peanuts, plums, cranberries, and cinnamon, but data about the molecular diversity of procyanidins are limited. These compounds possess potential health benefits as a result of their antioxidant, anticarcinogenic, and anti-inflammatory actions.

However, the best known bioactivity of A-type PACs is their capacity to inhibit the adhesion of pathogenic bacteria to uroepithelial cells of urinary tract, and therefore preventing infections. For this reason, in the last years several commercial products which claim to be manufactured from fruit-based extracts rich in A-type PACs have appeared in the market. However, some of these products do not contain the appropriate PACs for having the desired bioactivity. This fact shows the importance of developing analytical methodologies for the analysis and characterization of proanthocyanidins in natural extracts to achieve correct authentication regarding the fruit of origin.

This chapter will review the state-of-the-art of liquid chromatography-mass spectrometry techniques for the analysis and characterization of proanthocyanidins in natural products. LC-MS chromatographic conditions, ionization sources, and MS and HRMS analyzers frequently used, as well as strategies for the structural characterization and the qualitative and quantitative analysis of proanthocyanidins will be discussed by means of relevant applications. The use of chemometric tools to characterizing PAC polyphenolic profiles for the authentication of natural extracts will be addressed. Coverage of all kind of applications is beyond the scope of the present contribution, so the authors will focus on the most relevant applications published in the last years.

Chapter 2 - Oxidative stress results from pro-oxidant/antioxidant imbalance leading to an overproduction of reactive oxygen species (ROS) which can cause oxidative damage to cell structures and even trigger to cell death. Oxidative stress contributes to the progressive neurodegeneration associated to diseases such as Alzheimer´s disease, and Parkinson´s disease. The use of exogenous antioxidants presented in the diet that counteract oxidative stress damage by inhibiting free radical overproduction or by enhancement of the enzymatic and non-enzymatic defense mechanisms is being widely investigated as promising neuroprotector agents.

Proanthocyanidins (commonly known as condensed tannins) are secondary metabolites with oligomeric (dimers to pentamers) and polymeric (six or more units, up to 60) structures formed from the condensation of units of flavanols. These polyphenolic compounds are found in high proportions in cocoa products, seeds and skins of grapes, green and black tea, cinnamon and berries, among others. These compounds possess multiple therapeutic effects including anticancer, antimicrobial activity, cardioprotective role, prevention of atherosclerosis, and neuroprotective action. Many of these pharmacological benefits, including their neuroprotective action, are attributed to their significant antioxidant activity as free radical scavengers and as modulators of oxidant/antioxidant status.

The present chapter offers an overview of the health beneficial properties of proanthocyanidins in the prevention or reduction of ROS-induced oxidative damage in neurodegenerative diseases.

Chapter 3 - Procyanidins are polyphenolic compounds found in relatively high concentrations in fruits that have come to be known as biologically active in the context of promoting human health. Epidemiological evidence suggests that people who consume diets high in fruits and vegetables are less susceptible to inflammation-related diseases and diseases of multi-factorial pathogenesis such as metabolic syndrome, atherosclerosis, asthma, and cancer. Procyanidins are one of three types of proanthocyanidins and differ from the others based on the hydroxylation pattern of their B ring. They differ amongst themselves according to the type of linkage that connects the flavan-3-ol monomeric units. Despite progress, there is still a lack of measurement technologies to delineate the structure of higher polymers accurately, although specific dimer designations have been identified. Research is working toward understanding the health-promoting properties of procyanidins, specifically moving beyond their antioxidant capabilities. Bioavailability of designated dimers *in vivo* is not well understood, though there are data to suggest that *in vitro* experiments could be relevant in the human system. Population studies have shown an inverse relationship between procyanidin consumption and ill health, especially inflammatory conditions. Here the authors present a focused review of the properties of individual procyanidins and suspected mechanisms by which they could potentially regulate inflammation. Modulation of cytokine secretion, mechanisms involving signal transduction pathways - NF-κB, STAT, and MAPK, and procyanidin influence on the chemotaxis of inflammatory cells into the tissues, will all be discussed.

Chapter 4 - Neurodegenerative disorders such as Parkinson's disease (PD) and Alzheimer's disease (AD) represent increasing problems in our aging societies, mostly with an increased prevalence of both diseases with age. Evidence suggests that lifestyle factors, especially nutrition, can influence the incidence and onset of neurodegenerative diseases and thus a healthy diet is an essential factor for health aging. Anthocyanins are the largest group of water-soluble pigments in the plant kingdom and belong to the family of secondary metabolites known as flavonoids. Major sources of anthocyanins are blueberries, cherries, raspberries, strawberries, black currants, purple grapes, and red wine. Anthocyanin pigments and associated flavonoids have demonstrated ability to protect against a myriad of human diseases. In particular, several studies have shown that the consumption of diets rich in antioxidant components, such as anthocyanins, may modulate the progression of neurodegenerative diseases such as AD and PD. This chapter is aimed at presenting the neuroprotective effects of anthocyanins and their interactions with specific molecular targets and pathways of neurodegeneration process. Potentials and pitfalls of dietary anthocyanins as neuroprotective agents have been discussed as well.

Chapter 5 - Proanthocyanidins are a class of "bioactive" compounds found in a number of natural products and can be complex in their structure and in their composition. Proanthocyanidins have been found in grape seeds, blueberries, cranberries and apples amongst many other sources. This paper reviews the existing recent research on the anti-cancer properties of proanthocyanidins with a particular focus on the proanthocyanidins found in grape seeds, blueberries, cranberries, and apples. Proanthocyanidins have been proposed to exert their anti-cancer effects by either inhibiting cellular proliferation, by the induction of apoptosis, by affecting the inflammatory response or by inhibiting the invasive capabilities of cancer cells, amongst other possible mechanisms. These proanthocyanidin –mediated effects involve alterations in a number of cellular activities and also involve alterations in various cellular signal transduction mechanisms. Collectively, these findings suggest a potential role for proanthocyanidins found in specific natural products to possess possible chemo-preventative and/or possible chemo-protective properties against cancer cells and as such, supports the inclusion of these proantho-cyanidin-containing natural products in an "anti-cancer" diet.

Chapter 6 - Apple polyphenols (APs) mainly consist of procyanidins (PCs), which are composed of polymerized (–)-epicatechins and (+)-catechins. APs are contained in the edible part of the apple, so apples may be good source of dietary polyphenols. APs have been reported to display several beneficial health effects. In this chapter, the authors summarize the anti-aging effects of APs on age-related diseases in model organisms. First, treatment with APs or PCs purified from APs extended the lifespan of *Caenorhabditis elegans* to a similar extent as treatment with *trans*-resveratrol. In contrast, treatment with APs had no effect on the longevity of *sir-2.1* worms, which lack the activity of SIR2, a member of the sirtuin family, indicating that APs showed SIR2-dependent anti-aging effects. In general, polyphenols have anti-oxidant activity. APs and PC dimers showed a higher level of superoxide dismutase-like activity and oxygen radical absorbance capacity than other polyphenols or antioxidants, as well as anti-oxidant effects on mouse myoblast cells *in vitro*. Next, the authors orally administered APs to heart/muscle-specific manganese-superoxide dismutase (Mn-SOD)-deficient mice, which exhibit lethal dilated cardiomyopathy due to intrinsic enhanced mitochondrial superoxide. Dietary APs significantly suppressed the progression of cardiac fibrosis, which extended the survival of the mutant mice. Dietary APs also suppressed oxidative DNA damage and reactive oxygen species production in enzymatically-dissociated cardiomyocytes of the mutant mice. Notably, dietary APs also improved the electrophysiological abnormalities, such as action potential and inwardly rectifier potassium current changes in the cardiomyocytes of the mutant mice, as determined using patch clamp techniques, and decreased the incidence of ventricular tachycardia or the induction of fibrillation using a burst ventricular pacing protocol. Furthermore, APs remarkably suppressed amyloid-β protein 42 (Aβ42) aggregation and dissociated Aβ42 aggregates *in vitro*, indicating that APs are potent suppressors of Aβ aggregation. APs also significantly inhibited Aβ neurotoxicity and stimulated the proliferation of neural PC-12 cells. Interestingly, treatment with APs also reduced polyglutamine aggregation in the *C. elegans* Q40 strain, a model of Huntington's disease. These findings strongly indicate that APs showed anti-aging effects, such as anti-oxidative, anti-protein aggregation and pro-longevity effects, using model organisms. Apple procyanidins are promising functional foods to prevent age-related disorders.

In: Proanthocyanidins
Editor: Ingrid Sullivan

ISBN: 978-1-63463-430-4

Chapter 1

LIQUID CHROMATOGRAPHY-MASS SPECTROMETRY IN THE ANALYSIS AND CHARACTERIZATION OF PROANTHOCYANIDINS IN NATURAL PRODUCTS

***Meritxell Navarro*[1] *and Oscar Núñez*[2,*]**
[1]Department of Environmental Chemistry, IDAEA-CSIC, Barcelona, Spain
[2]Department of Analytical Chemistry, University of Barcelona, Barcelona, Spain

ABSTRACT

Proanthocyanidins (PACs), present in a wide variety of food products and beverages, are the second most abundant class of phenolic compounds in our diet. The most common PACs are procyanidins, exclusively consisting of (epi)catechin units. Besides procyanidins, prodelphinidins, a heterogeneous group consisting of at least one (epi)gallocatechin unit and additional (epi)catechin units, and propelargonidins, a heterogeneous group consisting of at least one (epi)afzelechin unit and additional (epi)catechin units, occur in food sources. PACs consist of monomeric flavan-3-ol units, which are mainly linked through C4-C8 or sometimes C4-C6 bonds. These linkages are

* Email address: oscar.nunez@ub.edu

both called B-type linkages. When an additional ether linkage is formed (mainly) between C2-O-C7, compounds are called A-type PACs.

The most well-known sources of A-type PACs are peanuts, plums, cranberries, and cinnamon, but data about the molecular diversity of procyanidins are limited. These compounds possess potential health benefits as a result of their antioxidant, anticarcinogenic, and anti-inflammatory actions. However, the best known bioactivity of A-type PACs is their capacity to inhibit the adhesion of pathogenic bacteria to uroepithelial cells of urinary tract, and therefore preventing infections. For this reason, in the last years several commercial products which claim to be manufactured from fruit-based extracts rich in A-type PACs have appeared in the market. However, some of these products do not contain the appropriate PACs for having the desired bioactivity. This fact shows the importance of developing analytical methodologies for the analysis and characterization of proanthocyanidins in natural extracts to achieve correct authentication regarding the fruit of origin.

This chapter will review the state-of-the-art of liquid chromatography-mass spectrometry techniques for the analysis and characterization of proanthocyanidins in natural products. LC-MS chromatographic conditions, ionization sources, and MS and HRMS analyzers frequently used, as well as strategies for the structural characterization and the qualitative and quantitative analysis of proanthocyanidins will be discussed by means of relevant applications. The use of chemometric tools to characterizing PAC polyphenolic profiles for the authentication of natural extracts will be addressed. Coverage of all kind of applications is beyond the scope of the present contribution, so we will focus on the most relevant applications published in the last years.

INTRODUCTION

Over the past years, researchers and food manufacturers, as well as the public in general, have become increasingly interested in polyphenolic compounds. These compounds are aromatic secondary metabolites that are ubiquitously spread through the plant kingdom. They comprise more than 8000 substances with highly diverse structures and molecular masses ranging from small molecules (<100 Da) such as phenolic compounds to big molecules (>30,000 Da) for highly polymerized polyphenolic compounds. The main reason for the interest in this family of compounds is the recognition of their antioxidant properties, their great abundance in our diet, and their probable role in the prevention of various diseases, such as skin pathologies, various types of cancer, cardiovascular disorders, and other age-related degenerative pathologies, besides the general health benefits they provide [1-6]. Furthermore, polyphenols, which also constitute the active substances found in many medicinal plants, modulate the activity of a wide range of enzymes and cell receptors [7]. Due to all these aspects, the importance of diet on human health and well-being has been widely recognized all over the world. For instance, in 1991, the United States launched "the 5 a day for Better Health Program" to increase consumption of vegetables and fruits to an average of 5-9 servings a day, and the "Dietary Guidelines for Americans" of USA recommends that people consume, at least, 2.5 cups of vegetables and 2 cups of fruits daily [8], which is based on a general 2000 kcal/day diet. In Europe, the traditional Mediterranean diet has formed the basis for food consumption during the past century, originally settled on Mediterranean agronomical, pastoral, and rural archetypes, and which includes a traditional daily consumption of fruits and vegetables. Moreover, polyphenols are an important constituent of fruit quality because of their contribution to taste, color and nutritional properties [9].

Polyphenols may be classified into different groups as a function of the number of phenol rings that they contain and the structural elements that bind these rings to one another. For instance, they can be divided in phenolic acids (hydroxybenzoic acids and hydroxycinnamic acids), flavonoids (flavonols, flavones, flavanols, flavanones, isoflavones, proanthocyanidins), stilbenes, and lignans [10]. It should be commented that for flavonoids, more than 5,000 diferrent kinds have been described [11].

Proanthocyanidins (PACs) are also known as oligoflavanoids and they comprise condenses polyphenols or condensed tannins. Apart from lignin, they represent the most abundant class of natural phenolic compounds in our diet. The PACs obtained their name from the characteristic oxidative depolymerization reaction in acidic medium, which yields colored anthocyanidins. The most common PACs are procyanidins, exclusively consisting of (epi)catechin units. Besides procyanidins, prodelphinidins, a heterogeneous group consisting of at least one (epi)gallocatechin unit and additional (epi)catechin units, and propelargonidins, a heterogeneous group consisting of at least one (epi)afzelechin unit and additional (epi)catechin units, occur in food sources. PACs consist of monomeric flavan-3-ol units, which are mainly linked through C4-C8 or sometimes C4-C6 bonds. These linkages are both called B-type linkages. When an additional ether linkage is formed (mainly) between C2-O-C7, compounds are called A-type PACs. As an example, Figure 1 shows the structure of a trimetric proanthocyanidin with both A-type and B-type linkages.

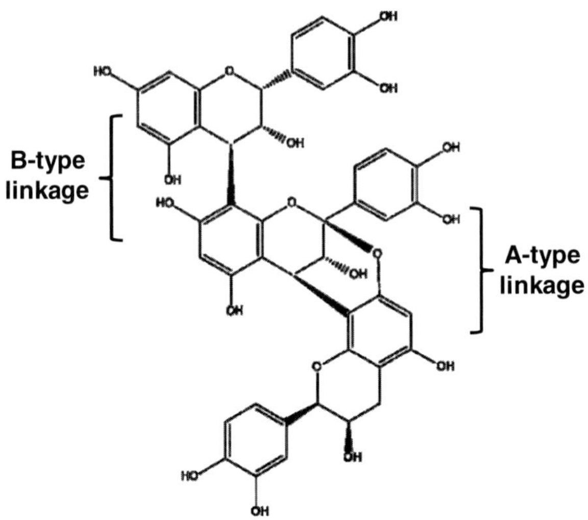

Figure 1. Representative structure of a trimeric proanthocyanidin with both A-type and B-type linkages. Reprinted with permission from reference [12]. Copyright (2014) American Chemical Society.

In addition to C2-O-C7 linkages, A-type PACs with C2-O-C5 linkages are also found, for instance in cocoa [13]. The flavan-3-ol units may carry also acyl or glycosyl substituents. The most common substituent bound as an ester is gallic acid to form 3-*O*-gallates. Several glycosylated PAC oligomers have also been identified, with the sugar linked to the C3 or the C5 position as the most widely distributed glycosylation.

Catechins and proanthocyanidins are found in common foods such as fruits (apples, pears, plums, strawberry, cranberry, kiwi, dates, many red fruits…), cereals (sorghum, barley...), seeds and nuts (beans, peas, almonds…), spices, aromatic plants, and more scarcely in vegetables [14]. They can also be found in various foodstuffs of plant origin (wines, tea, ciders, beers, chocolates, jams, puree…) [15]. However, in these processed foods, catechins and PACs are not only present in their native form, but they have sometimes undergone structural changes especially related to their susceptibility to oxidation with a significant impact on their physico-chemical properties. One of the most obvious examples is probably that of black tea catechins that are enzymatically oxidized, forming theaflavins and thearubigins responsible for the color of the infusions [16].

The extraction, separation, and analysis of catechins and proanthocyanidins are often hampered by several properties related to their structural diversity, their tanning properties, and also their sensitivity toward certain reactions such as oxidation [14].

In general, the elucidation of PAC structures has relied on extensive purification followed by spectral (NMR, MS) and chemical degradation with thiolytic or phloroglucional adduction. However, NMR resonance multiplicity and broadening associated with rotational and conformational isomerism often complicates the interpretation of their sequences [17]. Chemical degradation requires tedious large-scale separation, multiple purifications, and various chemical reactions, and at the end, many structural identities may remain unresolved for lack of authentic samples. Therefore, liquid chromatography coupled to mass spectrometry (LC-MS) has attracted greater attention due to its capacity for reliable analysis and small sample requirements. In this context, high resolution mass spectrometry (HRMS) is also playing an important role to help in the analysis and characterization of PACs in natural extracts.

ANALYSIS AND CHARACTERIZATION OF PACs BY LC-MS(/MS)

The analysis of proanthocyanidins, and among them of procyanidins, the most common group of PACs, is essential for a better understanding of whether naturally occurring concentrations of these compounds in food are sufficient for possible health effects.

Therefore, different attempts have been made to quantify PACs in food samples in the last years [18]. Several colorimetric methods, more or less selective, are available for the detection of catechins and PACs in plant tissues and extracts that may be also used for quantification, although their reliability of quantitative purpose is limited particularly in samples containing highly polymerized PACs.

For instance, phenolic compounds react with aromatic aldehydes in acidic conditions to yield colored products [19]. This is the base of the vanillin-HCl assay, in which vanillin or dimethylaminocinnamaldehyde (DMACA) as aldehydic reagent react with flavanols (catechins and PACs) leading to reddish and bluish colors, respectively. However, these assays allow only the determination of the total PAC content as a sum without a specification of individual compounds.

Normal phase liquid chromatography (NP-LC) on silica columns is a very useful technique for the separation of PACs, specially procyanidins, according to the degree of polymerization (DP) in clusters up to a DP of 10 [20-23].

Reversed phase liquid chromatography (RP-LC) is the most common method for the separation and analysis of phenolic compounds in plants, fruit products and beverages, and numerous works can be found in the literature [24-28]. It is also widely employed for the analysis of catechins and some PACs such as oligomeric procyanidins, but in most cases, it is not enough for the complete analysis and characterization of flavanols present in complex samples which contains a great variety of proanthocyanidins (isomers with different hydroxylation patterns) or polymerized tannins. Actually, RP-HPLC separation of oligomers larger than tetramers (DP of 4) is very difficult [14], and higher oligomeric and polymeric PACs will elute in an unresolved hump. The elution order will depend on the stereochemistry and overall polarity of single compounds; for example, C4-C8 linked dimeric PACs elute prior to the C4-C6 linked compounds.

For detecting PACs in complex extracts, typically time consuming methods are employed to pre-purify the extracts before separation and detection. After extraction and purification, PACs are commonly resolved by HPLC and detected either using UV-detection or mass spectrometry [29]. Mass spectrometry offers high selectivity by measuring the characteristic transitions of the different DPs found within PAC compounds.

Table 1 shows a selection of recently published LC-MS(/MS) methods for the analysis and characterization of PACs in food samples. As can be seen, from the chromatographic point of view the overall trend is to use conventional 3-5 μm particle size C18 RP columns [30-40] under binary gradient elution conditions using acidified water (with formic acid or acetic acid [35,37,38]), although sometimes other electrolytes such as ammonium acetate are used [36], and acetonitrile (ACN) [40] or methanol [36,37,39], or even acidified methanol [31,38] or ACN [30,32-35], as mobile phases. The use of ternary gradient elution conditions such as combinations with water, methanol and acetonitrile (acidified with acetic acid) mobile phases has also been reported for the separation of PACs according to their DP by using a Diol-phase column [35]. Other stationary phases such as diphenyl ones have also been reported for the conventional HPLC separation of PACs and phenolic compounds [41]. Recently, ultra-high performance liquid chromatography (UHPLC) using sub-2 μm particle size columns have also been described for the LC-MS(/MS) analysis and characterization of PACs in food products [42,43].

The use of other RP chromatographic phases, such as Phenyl columns, for the analysis of these compounds have also been described [43]. Regarding mass spectrometry, a real development in the analysis of PACs appeared in the late 1990s with the emergence of electrospray ionization (ESI) [44] and matrix-assisted laser desorption ionization (MALDI) techniques [45]. Since then ESI has been applied for the characterization of PACs in samples from various plant origins. ESI-MS analysis of catechins and PACs are generally performed in the negative ionization mode, although detection in positive ionization mode is also possible [32]. Triple quadrupole mass analyzers working in selected reaction monitoring (SRM) acquisition mode are among the most common MS instruments used for the LC-MS/MS analysis of PACs in food products, although other MS analyzers such ion trap [34-41] or linear ion trap [32] are also being employed, although more focused on characterization.

An interesting work is the one proposed by Rzeppa et al. [30] using a C18 reversed-phase liquid chromatography coupled to electrospray-tandem mass spectrometry (RP-LC-ESI-MS/MS) method for the analysis of Flavan-3-ols and procyanidins in food samples. In a first step, 12 procyanidins (B1, B2, B3, B4, B5, B6, B7, B8, C1, C2 and A2, and cinnamtannin B1) were isolated from plant materials or synthesized and purified by a combination of column chromatographic separation techniques with different stationary phases. Then, these procyanidins in combination with catechin and epicatechin were used as standard substances for the identification and quantification during the screening of food samples by the proposed RP-LC-ESI-MS/MS method. As an example, Figure 2 shows the chromatograms obtained for a 500 μg/L standard mixture (Figure 2a) and an apple sample (Figure 2b).

Acetone:water (70:30 *v:v*) solutions were used for the extraction of procyanidins from fruits after the removal of fat and chlorophyll with n-hexane. No further purification was carried out, but the chromatographic eluate during the first 9 minutes were derived to waste by means of a valve prior to the triple quadrupole MS/MS system to remove mainly sugars and other polar compounds. Identification and quantification was carried-out by MS using SRM acquisition mode, which offers the advantage of an unambiguous identification of procyanidins by measuring the characteristic mass spectrometric transitions of the different DPs in contrast to other detection methods, such as UV or fluorescence detection (transitions indicated in Figure 2 caption).

Table 1. Selection of LC-MS(/MS) methods for the analysis and characterization of proanthocyanidins in food samples

Compounds	Samples	Chromatographic conditions	Mass spectrometry conditions	Analysis time	LODs	Ref.
12 dimeric and trimeric PACs	55 fruit samples	LiChrospher C18 column (250x2 mm, 5 µm) *Gradient elution:* A) Water with 0.1% formic acid B) ACN with 0.1% formic acid 0.3 mL/min	Triple quadrupole mass analyzer (-) ESI SRM acquisition mode	35 min	-	[30]
Short oligomeric PACs with other polyphenols	Boysenberry seeds and juice	C18 reversed-phase column (150x2.1 mm, 3 µm) *Gradient elution:* A) Water with 0.1% formic acid B) Methanol with 0.1% formic acid 0.2 mL/min	Triple quadrupole mass analyzer (-) ESI SRM acquisition mode	50 min	-	[31]
Several monomeric, dimer, trimer and tetramer PACs	Green, oolong and black tea crude extracts, and grape seed extract.	Luna C18(2) column (150x2.1 mm, 5 µm) *Gradient elution:* A) Water with 0.1% formic acid B) ACN with 0.1% formic acid 0.3 mL/min	Linear ion trap (LTQ) mass analyzer (+) ESI Product ion scan (full scan MS^2)	35 min	-	[32]
Anthocyanidins and PACs (B1, B2, A2)	Wine lees	Mediterranea Sea C18 column (150x4.6 mm, 3 µm) *Gradient elution:* (A) Water with 0.5% formic acid (B) ACN with 0.5% formic acid 0.8 mL/min *Combined with on-line SPE extraction (Prospekt) with HySpere C8 EC cartridges*	Triple quadrupole mass analyzer (+) ESI for anthocyanidins (-) ESI for PACs SRM acquisition mode	18 min	15-30 ng/L	[33]

Compounds	Samples	Chromatographic conditions	Mass spectrometry conditions	Analysis time	LODs	Ref.
Profiling of PACs	Egiptian palm (*Medemia argun)*	C18 RP column (150x2.1 mm, 5 µm) *Gradient elution:* (A) Water with 0.03% formic acid (B) ACN with 0.03% formic acid 0.4 mL/min	Ion trap mass analyzer (-) ESI Product ion scan (full scan MS^2)	34 min	-	[34]
Phenolic acids and PACs	U.S. Pecans	Luna C18(2) column (250x4.6 mm, 5 µm) *Gradient elution:* A) Water:ACN:acetic acid (93:5:2 *v/v/v*) B) Water:ACN:acetic acid (58:40:2 *v/v/v*) 1.0 mL/min PrincetonSPHER DIOL column (250x4.6 nnm 5 µm) A) ACN:acetic acid (98:2 *v/v*) B) Methanol:ACN:acetic acid (95:3:2 *v/v/v*) 1.0 mL/min	Ion trap mass analyzer (-) ESI Product ion scan (full scan MS^2)	25 min	-	[35]
PACs	Virginia type peanut skins	Atlantis dC18 column (150x2.1 mm, 3 µm) *Gradient elution:* A) Water with 5 mM ammonium acetate B) Methanol 0.15 mL/min	Ion trap mass analyzer (-) ESI Product ion scan (full scan MS^2)	80 min	-	[36]
PACs	Pea, lentil and faba bean seeds	Symmetry C18 column (250x4.6 mm, 5 µm) *Gradient elution:* A) Water with 1% acetic acid B) Methanol 1.0 mL/min	Q-TRAP (quadrupole-ion trap) mass analyzer (-) ESI Product ion scan (full scan MS^2)	14 min	-	[37]

Table 1. (Continues)

Compounds	Samples	Chromatographic conditions	Mass spectrometry conditions	Analysis time	LODs	Ref.
PACs	African plant *Parkia biglobosa* (Fabaceae) roots and barks	Synergi Hydro C18 column (250x4.6 mm, 4 μm) *Gradient elution:* A) Water with 0.1% acetic acid B) Methanol with 0.1% acetic acid 0.8 mL/min	Ion trap mass analyzer (-) ESI Full scan MS	60 min	-	[38]
PACs and other phenolics	Boldo Folium (*Peumus boldus Mol.)* infusions	Luna C18 column (250x4.6 mm, 5 μm) *Gradient elution:* A) Water with 1% formic acid B) Methanol 1.0 mL/min	Ion trap mass analyzer (-) ESI Full scan MS	90 min	-	[39]
PAC A2	Concentrated cranberry juice	Zorbax SB-C18 (15x1 mm, 5 μm) *Gradient elution:* A) Water with 0.1% formic acid B) ACN 0.1 mL/min	Ion trap mass analyzer (-) ESI Full scan MS	14 min	-	[40]
PACs and other phenolics	Plum fruits (*Prunus salicina L.* and *Prunus domestica L.*)	Diphenyl column (150x3 mm, 5μm) *Gradient elution:* A) Water with 0.005% formic acid B) Methanol 0.5 mL/min	Ion trap mass analyzer (-) ESI MS^n fragmentation	160 min	-	[41]

Compounds	Samples	Chromatographic conditions	Mass spectrometry conditions	Analysis time	LODs	Ref.
Several PACs together with flavonol glycosides, terpene lactones, biflavones, and Ginkgolic acids	*Ginkgo biloba* Leaves from Fruit Cultivars	Acquity UPLC BEH C18 column (100x2.1 mm, 1.7 μm) *Gradient elution:* A) Water with 0.1% formic acid B) ACN 0.4 mL/min	Triple quadrupole mass analyzer (-) ESI SRM acquisition mode	16 min	0.04-3.5 μg/L	[42]
PAC oligomers	Crude plant extracts	Acquity UPLC BEH Phenyl column (100x2.1 mm, 1.7 μm) *Gradient elution:* A) Water with 0.1% formic acid B) ACN 0.5 mL/min	Triple quadrupole mass analyzer (-) ESI SRM acquisition mode	6 min	16 μg/L for procyanid in dimers	[43]

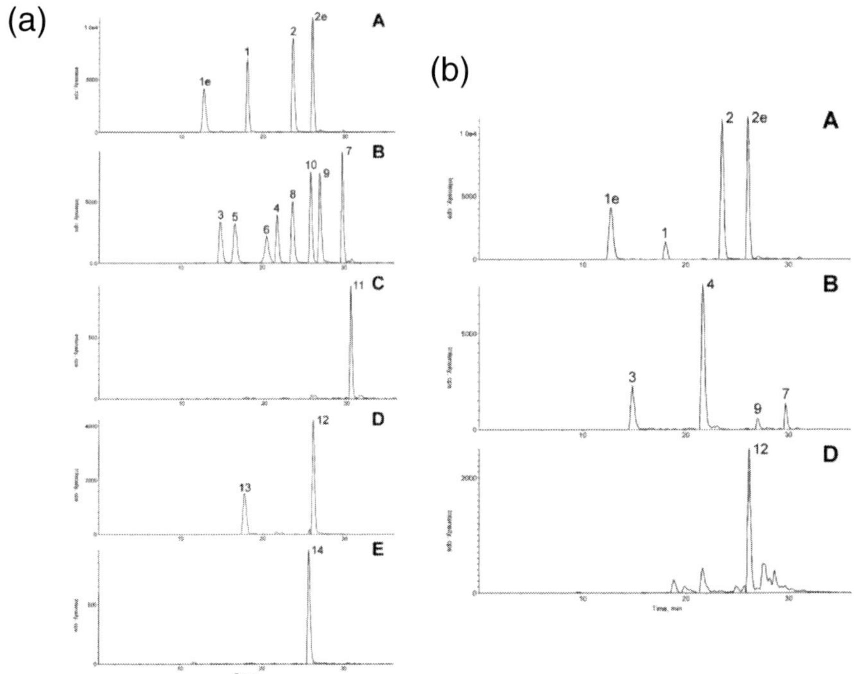

Figure 2. RP-HPLC-ESI-MS/MS chromatograms of (a) a mixture of standard substances (500 µg/L) and (b) an apple sample (Braeburn). (A) monomeric compounds with SRM *m/z* 289.0-124.9 $[M-H]^-$; (B) dimeric procyanidins with SRM *m/z* 577.1-124.9 $[M-H]^-$; (C) dimeric procyanidins A-type with SRM *m/z* 575.0-284.9 $[M-H]^-$; (D) trimeric procyanidins with SRM *m/z* 865.3-125.1 $[M-H]^-$; (E) trimeric procyanidins A-type with SRM *m/z* 863.1-284.9 $[M-H]^-$. Peaks: cathecin-echo (1e); catechin (1); epicatechin (2); epicatechin-echo (2e); procyanidins B1 (3), B3 (5), B4 (6), B2 (4), B6 (8), B8 (10), B7 (9), B5 (7), A2 (11), C2 (13), and C1 (12); cinnamtannin B1 (14). Reprinted with permission from reference [30]. Copyright (2011) American Chemical Society.

The difficulty of an accurate quantification of PACs due to the appearance of matrix effects when electrospray and MS techniques are used is illustrated in the work reported by Rzeppa et al. [30]. In complex food samples, analyte signals can be suppressed or enhanced in the presence of matrix compounds. The most usual solution will be the use of stable isotope labeled substances as internal standard, but in the case of PAC analysis, the synthesis of all PAC standards as isotopically labeled standards is very complex and time-consuming. So, for quantification purposes, and in order to take into account the matrix effects, the authors proposed an interesting, but until now rarely used, solution based on the so-called echo-peak calibration techniques [46] with a time-shifted injection of catechin and epicatechin solutions of known concentrations. The principle of this echo-peak technique in the analysis of PACs is shown in Figure 2. Solutions of catechin and epicatechin in a constant concentration were injected time-shifted to the calibration standard or sample. Catechin eco-peak (peak 1e) was injected before injection of the calibration standard or sample, and epicatechin eco-peak (peak 2e) after the injection, so a total of three injections are performed for each chromatographic run. Then the eco-peaks are used as internal standards during quantification. However, the authors did not carried-out the quantification of every studied analyte by using its own echo-peak because of the complexity of the PAC separation as well as the presence of overlapping MS transitions. The authors applied the proposed method to the analysis of more than 50 fruit samples, reporting concentrations levels between 0.1 mg/Kg to up to 150 mg/Kg (for instance procyanidin B2 in Elstar apples).

Another interesting approach for the quantitative analysis of PAC oligomers and polymers is the one recently described by Engström et al. using an UHPLC-MS/MS method [43]. The method proposed by the authors utilizes a range of cone voltages to achieve the depolymerization of both smaller PAC oligomers and larger polymers in the ion source. The formed depolymerization products are further fragmented in the collision cell of a triple quadrupole mass analyzer to enable their selective detection. This UHPLC-MS/MS method is able to separately quantificate the terminal and extension units of the most common PAC subclasses, i.e., procyanidins and prodelphinidins. As an example, mass spectra of PAC monomers, dimers, trimers, and polymers obtained by in-source fragmentation are shown in Figure 3a, and the SRM chromatograms of procyanidin and prodelphinidin extension and terminal units, respectively, for two plants leaves samples, in Figure 3b. The resulting data enable (i) the quantification of the total PAC content, (ii) the quantification of total procyanidins and prodelphinidins including the procyanidin/prodelphinidin ratio, (iii) the estimation of the mean degree of polymerization for the oligomers and polymers, and (iv) the estimation of how the different procyanidin and prodelphinidin types are distributed along the chromatographic hump typically produced by large PACs, and all this achieved within the 10 min period of the UHPLC-MS/MS analysis. For instance, to determine the total content of PACs and procyanidin /prodelphinidin ratios, the authors obtained calibration curves with the developed SRM method using pure procyanidins and pure prodelphinidins.

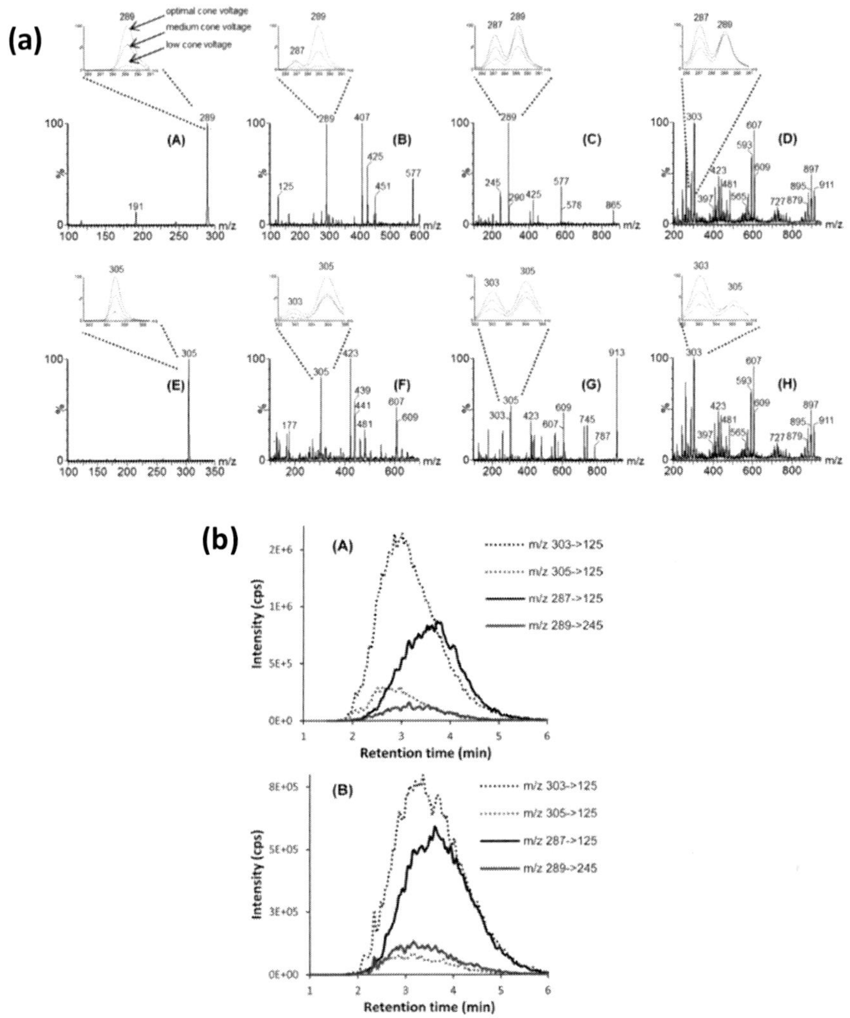

Figure 3. (a) Mass spectra of PAC monomers, dimers, trimers, and polymers obtained by in-source fragmentation: (A) catechin; (B) procyanidin dimer; (C) procyanidin trimer; (D) and (H) *Onobrychis viciifolia* (sainfoin: Nova var.), leaves sample; (E) epigallocatechin; (F) prodelphinidin dimer; and (G) prodelphinidin trimer. The insets show how the intensities of the extension units (*m/z* 287 and 303) and terminal units (*m/z* 289 and 305) change due to shifts in cone voltage. (b) SRM chroamtograms of procyanidin extension units (*m/z* 287 → 125), procyanidin terminal units (*m/z* 289 → 245), prodelphinidin extension units (*m/z* 303 → 125), and prodelphinidin terminal units (*m/z* 305 → 125) for (A) *Onobrychis viciifolia* (sainfoin: Nova var.) and (B) *Lotus corniculatus* (birdsfood trefoil: Rodeo var.) leave samples. Reprinted with permission from reference [43]. Copyright (2014) American Chemical Society.

Five cone voltages were used for detection, and resulting peak areas were summed. These calibration curves were then used for estimation of procyanidin and prodelphinidin concentrations in 18 PAC fractions from plant leaves samples. The authors also compared the resulting PAC contents with those estimated by thiolysis and HCl-butanol assay. Good correlation for PAC contents with both thyolysis (R^2 = 0.82) and HCl-butanol (R^2 = 0.86) was achieved with the proposed UHPLC-MS/MS method.

Although some publications are describing approaches for the quantification of PAC compounds in food products, as previously commented, most of the works reported in the literature focus on the identification and characterization of PACs by using LC-MS(/MS) [31,34,35]. For instance, Robbins et al. proposed a LC-ESI-MS/MS method using an ion-trap mass analyzer for the separation and characterization of phenolic acids and PACs from U.S. pecans [35]. The authors selected eight commercially significant cultivars based on their relatively high antioxidant capacities to obtain crude phenolic extracts. These pooled extracts were then separated via Sephadex LH-20 chromatographic column into five ethanolic low-molecular weight (LMW) fractions and one acetonic high-molecular-weight (HMW) fraction, and characterized using LC-ESI-MS/MS. Figure 4 shows, as an example, the diol-phase HPLC chromatogram for the tannin fraction isolated from the eight pooled pecan cultivars. As can be seen, this HMW fraction contained most of the PACs, basically as dimers, but monomers, trimers, tetramers, pentamers, and even hexamers were also separated by using a diol-phase chromatographic column. A deeper examination of the PAC composition in U.S. peanuts was completed by the authors using diol-phase HPLC-ESI-MS/MS analyses, and several PAC compounds were identified in the acetonic fraction (see Table 2). All of the compounds present in the sample exhibited a B-type linkage, which was expected, based on the literature reports regarding pecan PACs [15].

Interestingly, the tannin fraction studied by the authors possessed both procyanidin and prodelphinidin compounds. Two PACs, one procyanidin and one prodelphinidin, were detected for each DP with the exception of the dimer, which had a single identification, and the hexamers, which were both prodelphinidins. Typically, ion-trap mass analyzers are frequently used for the characterization of PACs in food products because their strong capacities to perform MS^n fragmentations, allowing the proposal of PAC fragmentation pathways, very useful for the identification of these compounds in complex matrices. For example, Jin et al. [37] proposed an LC-ESI-MS/MS method using and ion-trap mass analyzer for the characterization of proanthocyanidins in pea (*Pisum sativum L.*), lentil (*Lens culinaris L.*) and faba bean

(*Vicia faba L.*) seeds, and Figure 5 shows the proposed fragmentation pattern for gallocatechin-(4α→2)-phloroglucinol PAC.

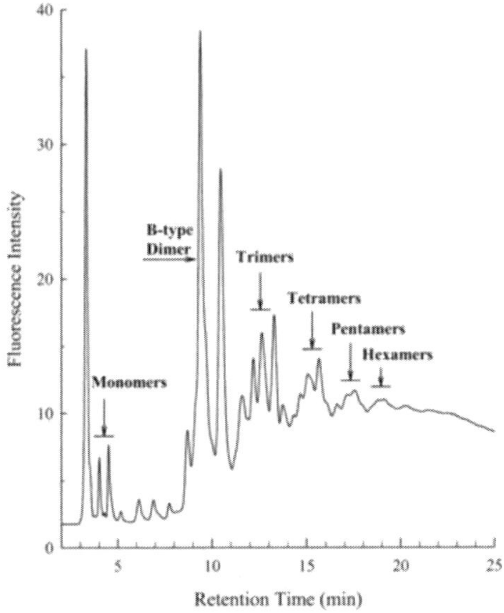

Figure 4. Diol-phase HPLC chromatograms of the tannin fraction isolated from eight pooled pecan cultivars. Fluorescence detection was employed with excitation/emission at λ = 276/316 nm, respectively. Reprinted with permission from reference [35]. Copyright (2014) American Chemical Society.

Table 2. PAC compounds present in the Tannin Fraction isolated from eight pooled pecan cultivars

t_r (min)[a]	DP	unit type[b]	linkage	$[M-H]^-$
6.5	2	2 (epi)catechin	B	577
9.7	3	3 (epi)catechin	B	865
11.0	3	2 (epi)catechin and 1 (epi)gallocatechin	B	881
13.1	4	4 (epi)catechin	B	1153
14.2	4	3 (epi)catechin and 1 (epi)gallocatechin	B	1169
16.0	5	5 (epi)catechin	B	1441
16.3	5	5 (epi)catechin and 1 (epi)gallocatechin	B	1457
18.0	6	4 (epi)catechin and 1 (epi)gallocatechin	B	1745
19.0	6	4 (epi)catechin and 2 (epi)gallocatechin	B	1761

[a] Retention time determined from diol-phase HPLC-ESI-MS analysis.

[b] Cross-checking with compounds identified in nut skins, as reported by Sarnoski et al. [36] and Monagas et al. [47].

Table reproduced with permission from reference [35]. Copyright (2014) American Chemical Society.

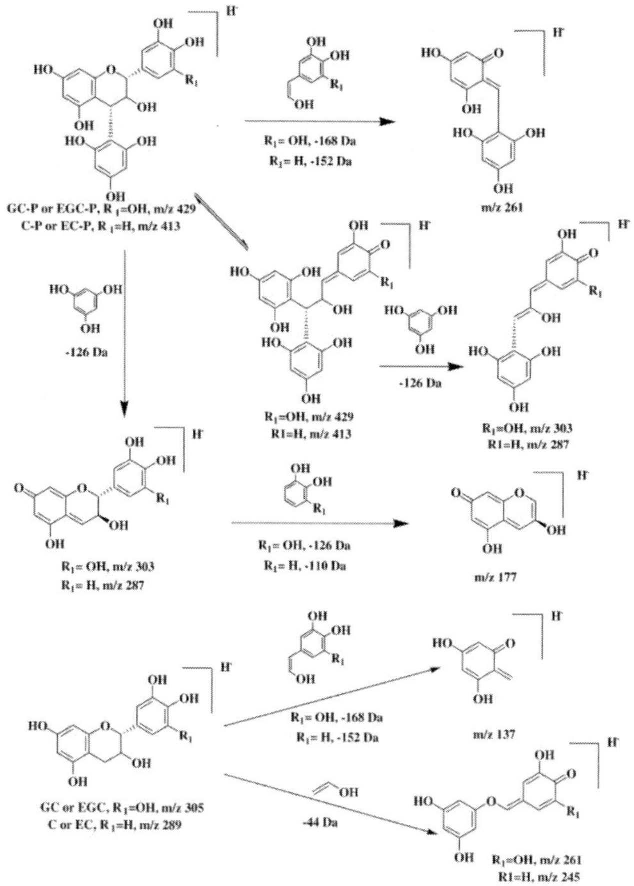

Figure 5. Proposed MS fragmentation patterns of GC-P (gallocatechin-(4α→2)-phloroglucinol). GC, gallocatechin; EGC-P, epigallocatechin-(4β→2)-phloroglucinol; EGC, epigallocatechin; C-P, catechin-(4α→2)-phloroglucinol; C, catechin; EC-P, epicatechin-(4β→2)-phloroglucinol; EC, epicatechin. Modified by the authors from references [40,48]. Reprinted with permission from reference [37]. Copyright (2012) Elsevier.

PAC subunit composition, quantitation and DP were determined by the authors after acid-catalysis in presence of excess phloroglucinol (phloroglucinolysis), and LC-ESI-MS/MS was used to confirm PAC subunit identity. B-type PACs were found in all three grain legume species. In the pea cultivars, PAC subunit composition was almost exclusively prodelphinidin (gallocatechin and epigallocatechin). In contrast, both procyanidin and prodelphinidin subunits were abundant in the PACs of faba bean and lentil seeds. PAC contents in the evaluated seeds ranged from approximately 270 to 350 mg/100 g fresh weight.

Two dimensional chromatography coupled to tandem mass spectrometry has also been reported for the characterization of PACs. For instance, recently Montero et al. [49] developed and optimized a comprehensive two-dimensional chromatography method involving the use of a microbore hydrophilic interaction liquid chromatography (HILIC) column with a diol stationary phase in the first dimension coupled to either a C18 partially porous short column or a C18 monolithic column in the second dimension. This orthogonal HILIC-RP-LC system was interfaced through a ten-port two-position switching valve, and coupled to diode array (DAD) and tandem mass spectrometry detection. A complex grape seed extract was analyzed by the proposed HILICxRP-LC-DAD-MS/MS method using ESI and an ion trap mass analyzer, allowing the identification of 43 flavan-3-ols, including monomers and procyanidin oligomers till a polymerization degree of 7 units with different galloylation degrees.

HIGH RESOLUTION MASS SPECTROMETRY

As previously commented, the emergence in the 1990s of ESI and MALDI represented an important development in the analysis of PACs. Since then, MALDI coupled to time-of-flight (TOF) high resolution mass spectrometry (HRMS) instruments have been applied for the characterization of proanthocyanidins in samples from various plant origins, as recently reviewed [50]. MALDI is a soft ionization energy method with a high ion transmission yield and with a great tolerance to the presence of impurities in the samples. However, it is not really adapted for the analysis of small catechin molecules (lower than 500 Da) since this region of the spectrum is generally saturated with matrix signals. MALDI-TOF analyses have been successfully applied for the characterization of PAC structures in a variety of fruits and plant extracts [51-55]. For instance, recently Chen et al. [55] characterized the structure of PACs from *Caryota ochlandra* Hance by combining ^{13}C-nuclear magnetic resonance (NMR), HPLC-ESI-MS/MS and MALTI-TOF-MS techniques. Although the ^{13}C-NMR and RP-HPLC-ESI-MS/MS spectra revealed complex structural characteristics of the fruit pericarp and flesh PACs, as can be seen, as an example, in Figure 6, quantitative data regarding the high degree of polymerization for these PACs could not be reliable obtained. Further characterization was achieved by MALDI-TOF-MS.

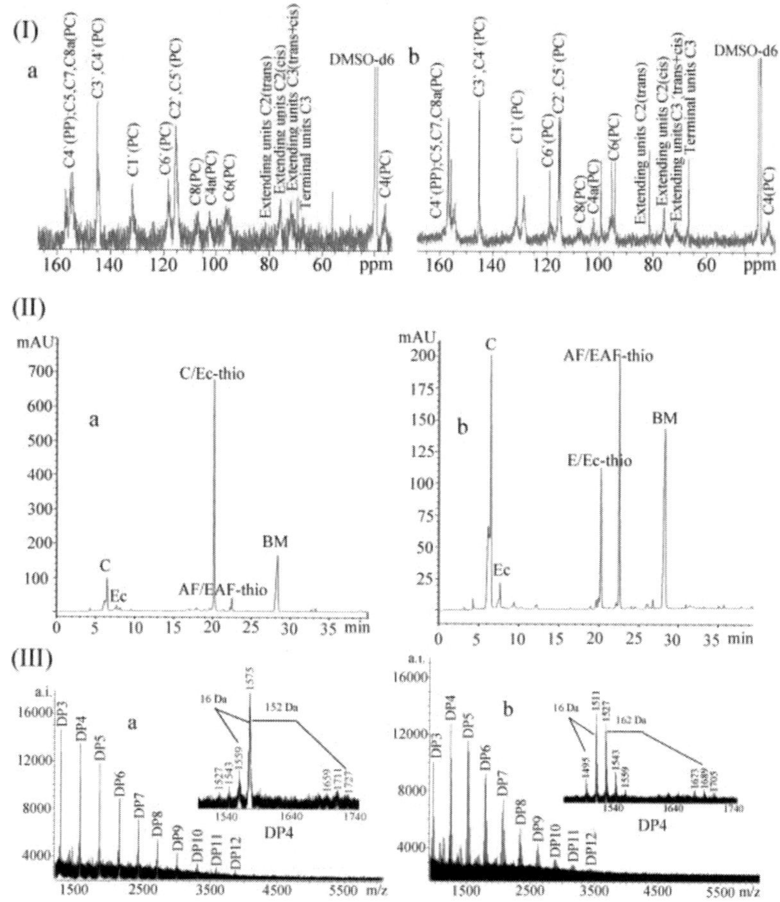

Figure 6. Structure characterization of proanthocyanidins. (I) ^{13}C-NMR spectroscopy. PC, procyanidin; PP, propelargonidin. (II) RP-HPLC-ESI-MS chromatograms. C/EC terminal unit catechin/epicatechin; C-thio/EC-thio and AF-thio/EAF-thio represented extender unit catechin/epicatechin and afzelechin/epiafzelechin, respectively; BM, represented benzyl mercaptan. (III) MALDI-TOF-MS spectra of the PACs. DP, degree of polymerization. (a) and (b) represented the fruit pericarp and flesh PACs of *Caryota ochlandra* Hance, respectively. Reprinted with permission from reference [55]. Copyright (2014) Elsevier.

Figure 6-III shows the MALDI-TOF-MS spectra of the polymeric PACs isolated from the fruit pericarp and flesh recorded as Catechin cation adducts in the positive ion reflectron mode and showing repeating PAC polymers. For both, fruit pericarp and flesh, PACs were detected up to dodecamer. The polydisperse PACs were reflected by the periodic occurrence of peak series representing different chain length. The results obtained indicated that the fruit pericarp and flesh PACs of *Caryota ochlandra* Hance where homo- and heteropolymers.

But today, ESI ionization techniques coupled to high resolution mass spectrometry is becoming also a very powerful tool for the analysis and characterization of PACs in food samples, and a selection of recently published ESI-HRMS methods for this purpose is summarized in Table 3. Some PAC characterization studies have been carried out with the direct injection of extracts into the ESI-HRMS system [56,57]. For instance, Touriño et al. [56] identified polyphenols, including PACs, from a grape antioxidant dietary fiber by the direct infusion of a polymeric fraction extract dissolved in formic acid:water:ACN (5:15:80 *v/v/v*) into an ESI-TOF-MS instrument. In contrast, Mouls et al. [57] characterized condensed tannins from cider apple cultivars by flow injection analysis (FIA) using water:ACN (50:50 *v/v*) coupled to ESI-TOF-MS. But in general, the characterization of PACs in food products is being carried out by HPLC or UHPLC techniques coupled to HRMS analyzer, either using time-of-flight instruments [58-62] or the most recently developed Orbitrap MS analyzers [63-65].

Similarly to conventional LC-ESI-MS(/MS) methods, reversed-phase columns with C18 stationary phases are usually employed when dealing with the analysis and characterization of PACs by LC-ESI-HRMS techniques, using acidified water and ACN (acidified or not) as mobile phases under gradient elution conditions. Negative ESI is the preferred ionization mode for the analysis of PACs, although in some cases positive ESI is also employed [57,61,66]. Among the available HRMS analyzers able to perform accurate mass measurements, TOF MS instruments [57-60], or the hybrids quadrupole-TOF (Q-TOF) instruments [61,62,66], are the most frequently employed when dealing with the characterization and analysis of PACs in natural products.

For instance, Iswaldi et al. [59] reported the characterization by LC-DAD-ESI-TOF-MS of the phenolic fraction in a cranberry syrup used to prevent urinary tract diseases. Both negative and positive ESI ionization modes were employed, and, as an example,

Figure 7 shows the base peak chromatographic profiles obtained with both ionization modes. The LC eluent flow delivered into the TOF-MS system was split using a flow splitter 81:2) to achieve stable electrospray ionization and obtain reproducible results.

Table 3. Selection of ESI-HRMS methods for the analysis and characterization of proanthocyanidins in food samples

Compounds	Samples	LC conditions	Mass spectrometry conditions	Analysis time	LODs	Ref.
PACs and other polyphenols	Grape antioxidant dietary fiber	Direct infusion of polymeric fraction dissolved in formic acid:water:ACN (5:15:80 *v/v*)	Q-TOF mass analyzer Mass resolution: 10,000 FWHM (-) ESI and (+) ESI	-	-	[56]
PACs and other polyphenols	Durum wheat	Zorbax C18 column (150x4.6 mm, 1.8 μm) *Gradient elution:* A) Water with 0.5% acetic acid B) ACN 0.125 mL/min	TOF mass analyzer (-) ESI	55 min	-	[58]
Condensed tannins	Cider apple cultivars	Flow injection analysis in water:ACN (50:50 *v/v*) 0.25 mL/min	TOF mass analyzer (-) ESI and (+) ESI	-	-	[57]
PACs and other polyphenols	cranberry syrup	Zorbax Eclipse Plus C18 column (150x4.6 mm, 1.8 μm) *Gradient elution:* A) Water:ACN (90:10 *v/v*) with 1% formic acid B) ACN 0.5 mL/min	TOF mass analyzer (-) ESI and (+) ESI LC eluent flow splitted (1:2) before MS system	30 min	150 μg/L for procyanidin A2	[59]
PACs	Czech hops	Acquity UPLC BEH Shield C18 column (50x2.1 mm, 1.7 μm) *Gradient elution:* A) Water with 0.1% formic acid B) ACN with 0.1% formic acid 0.4 mL/min	TOF mass analyzer (-) ESI	6 min	-	[60]
PACs and other polyphenols	Bark decoction of Saraca asoca	Zorbax C18 column (250x4.5 mm, 5 μm) *Gradient elution:* A) Water with 0.1% formic acid B) ACN with 0.1% formic acid 0.2 mL/min	Q-TOF mass analyzer (+) ESI	44 min	-	[61]

Table 3. (Continued)

Compounds	Samples	LC conditions	Mass spectrometry conditions	Analysis time	LODs	Ref.
PACs and other polyphenols	Apples	LiChroCART C18 column (250x4 mm, 4.5 μm) *Gradient elution:* A) Water with 1% formic acid B) ACN 0.8 mL/min	Q-TOF mass analyzer (-) ESI	40 min	-	[62]
PACs and other polyphenols	Diploid strawberry (*Fragaria vesca*)	Hypersil Gold C18 column (200x2.1 mm, 1.9 μm) *Gradient elution:* A) Water with 0.1% formic acid B) ACN with 0.1% formic acid 0.3 mL/min	Linear ion trap – Orbitrap (LTQ-Orbitrap) mass analyzer (-) ESI Tandem high resolution mass spectrometry ($HRMS^n$) Mass resolution: 100,000 FWHM	45 min	-	[63]
PACs and other polyphenols	Walnuts	Atlantis T3 C18 column (100x2.1 mm, 3 μm) *Gradient elution:* A) Water with 0.1% formic acid B) ACN with 0.1% formic acid 0.35 mL/min	Linear ion trap – Orbitrap (LTQ-Orbitrap) mass analyzer (-) ESI Full-scan, mass resolution: 60,000 FWHM Data dependent MS/MS, mass resolution: 30,000 FWHM	30 min	-	[64]
PACs and other polyphenols	Black, Green and White tea extracts	Luna C18(2) column (150x2.0 mm, 3μm) *Gradient elution:* A) Water with 0.1% formic acid B) ACN with 0.1% formic acid 0.19 mL/min	Linear ion trap – Orbitrap (LTQ-Orbitrap) mass analyzer (-) ESI Tandem high resolution mass spectrometry ($HRMS^n$)	45 min	-	[65]

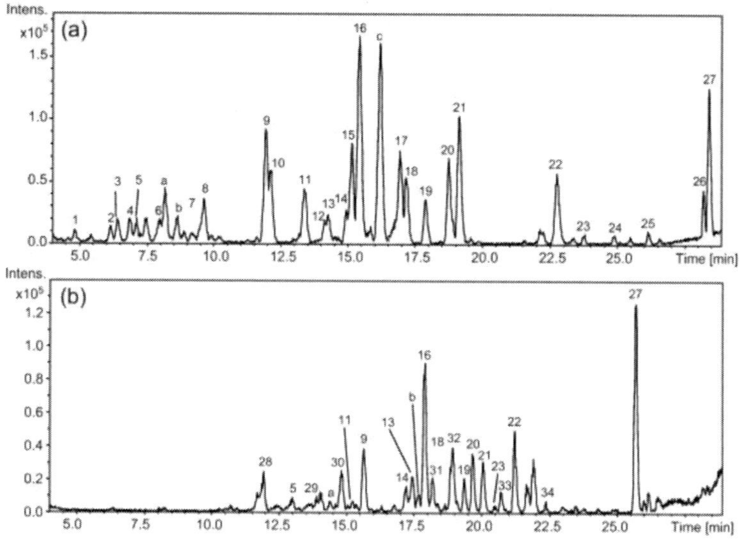

Figure 7. Base peak chromatographic profiles obtained from a Cranberry syrup by (a) LC-(-)ESI-TOF-MS and (b) LC-(+)ESI-TOF-MS. Reprinted with permission from reference [59]. Copyright (2012) Elsevier.

Regarding PACs and (-)ESI-TOF-MS analysis, three different isomers of A-type procyanidin were identified by the authors in the cranberry syrup sample at times 13.35, 14.28 and 16.94 min with a *m/z* of 575.121 (peaks 11, 13 and 17 in Figure 7a), showing the typical fragment at *m/z* 423.0 [59]. Two of these A-type procyanidin isomers were also detected with (+)ESI-TOF-MS (peaks 11 and 13 in Figure 7b, with *m/z* at 577.13). Two B-type procyanidin isomers with a *m/z* of 577.13 eluted at 8.00 and 14.14 min (peaks 6 and 12 in Figure 7a) and their fragments at *m/z* 425.0761 and 289.0687 were also detected. Peak 8 in Figure 7a, with a *m/z* of 289.0714 and a retention time of 9.65 min was identified as (+)-catechin. Prodelhpinidin B4 was also observed at a retention time of 26.11 min with a *m/z* of 609.1238 (peak 25 in Figure 7a). Accurate mass measurements with relative errors in the range 0.2-5.5 ppm were obtained. In general, mass resolutions around 10,000 full width at half maximum (FWHM) are achieved with TOF technologies, which is frequently enough for the analysis of characterization of PACs. But today, mass resolutions up to 150,000 FWHM are attainable with new Orbitrap HRMS technology [67,68], allowing a considerable improvement in accurate mass measurements (relative errors generally lower than 1-2 ppm). Moreover, the use of HPLC coupled to high-resolution tandem mass spectrometry (HR-MS/MS) can provide abundant information for structural elucidation of a wide range of compounds. Recently, the combination of Orbitrap technology with a liner ion trap (LTQ) (Thermo Fisher Scientific, San Jose, CA, USA) has been shown to enable fast, sensitive and reliable detection and identification of small molecules regardless of relative ion abundance [69]. This hybrid LTQ-Orbitrap HRMS analysis platform can easily perform multi-stage mass analysis (HRMS^n) and provide high mass accuracy measurements of precursor and product ions with fast scan speeds, becoming an important tool in the analysis and characterization of PACs among other polyphenolic compounds [63-65]. Recently, Regueiro et al. reported a comprehensive identification of walnut polyphenols by liquid chromatography coupled to linear ion trap-Orbitrap mass spectrometry [64]. A total of 120 compounds, including hydrolysable and condensed tannins, flavonoids and phenolic acids were identified or tentatively identified by the authors on the base of their retention times, accurate mass measurements (with errors lower than 3.18 ppm), and subsequent mass fragmentation data in HRMS, or by comparing with reference substances and literature. As an example, Figure 8 is showing the HRMS/MS fragmentation spectrum of a condensed tannin corresponding to a malabathrin A isomer, which was reported for the first time, among other seven polyphenols, in walnuts. Recently, comprehensive two-dimensional liquid chromatography coupled to high resolution mass spectrometry was exploited as a powerful analytical approach for the detailed characterization of grape seed tannins [70]. In this contribution, HILIC and reversed-phase chromatography were employed in the two dimensions to provide complementary information in terms of separation according to hydrophilicity and hydrophobicity, respectively.

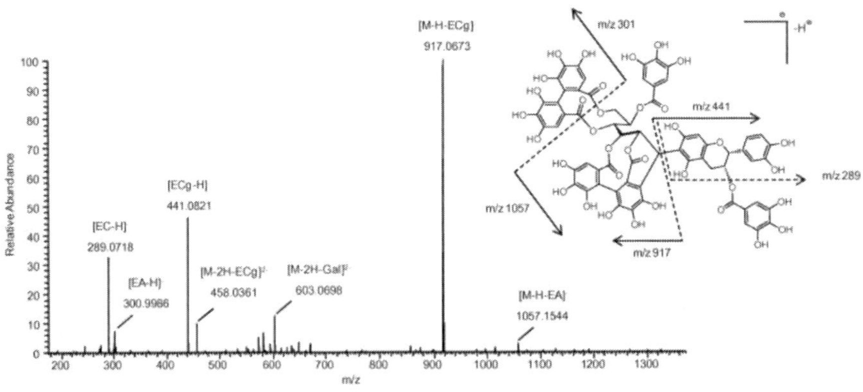

Figure 8. HRMS/MS fragmentation spectrum of a malabathrin A isomer. Reprinted with permission from reference [64]. Copyright (2014) Elsevier.

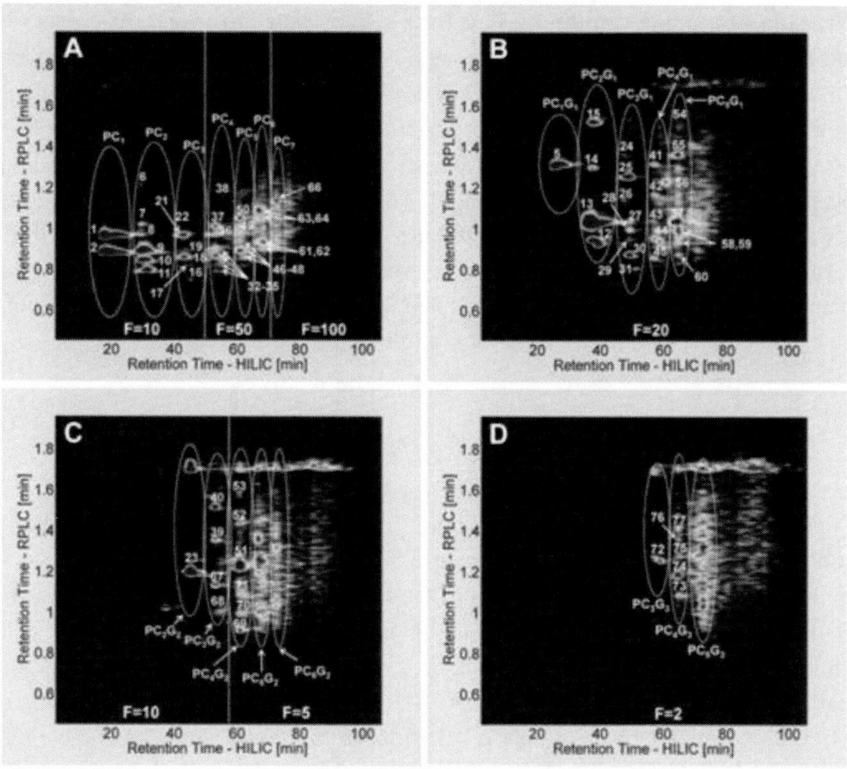

Figure 9. Extracted ion contour plots of (A) procyanidins, (B) monogalloylated procyanidins, (D) digalloylated procyanidins, and (D) trigalloylated procyanidins obtained by HILICxRP-LC-TOF-MS analysis of a red grape seed extract. F indicates the scaling factor used in the z-axis. In the notation PC_xG_y, *x* represents the procyanidin degree of polymerization and *y* represents the degree of galloylation. Reprinted with permission from reference [70]. Copyright (2013) American Chemical Society.

The HILICxRP-LC system was coupled with fluorescence detection and ESI-TOF-MS allowing selective detection and facilitating compound identification. Figure 9 shows the extracted ion contour plots of (A) procyanidins, (B) monogalloylated procyanidins, (D) digalloylated procyanidins, and (D) trigalloylated procyanidins obtained by by the proposed HILICxRP-LC-TOF-MS method in the analysis of a red grape seed extract. TOF-MS provided high acquisition rates and sensitivity coupled to accurate mass information.

A total of 78 PACs were identified as single chromatographic peaks, corresponding to the numbered compounds in Figure 9. These included 33 procyanidins, 25 monogalloylated, 11 digalloylated, 6 trigalloylated, and 1 tetragalloylated procyanidins, as well as 2 (epi)catechin monoglycosides. In addition, the high MS sensitivity achieved with the TOF-MS instrument allowed identification of low level procyanidins of DP up to 12 as well as monogalloylated-octagalloylated procyanidins of DP as high as 16. Other highly polymerized compounds were also identified, but isomeric separation in the RP-LC dimension was not possible. This methodology represents the most powerful single analytical method for the characterization of grape seed PAC chemical composition reported to date. And, given the importance of PACs in many natural products, this approach shows promise as a means of addressing the current analytical limitations for complex tannin characterization.

Chemometric Techniques

Analytical data obtained from liquid chromatography-mass spectrometry techniques contains huge amounts of information about analyzed samples. Treating this data by means of multivariate statistical analysis increases the quality of results and facilitates its interpretation, and therefore data information can be better exploited. Chemometric tools most used for characterizing PAC polyphenolic profiles for the authentication of natural extracts on the recent works are principal component analysis (PCA) and hierarchical cluster analysis (HCA) [71,72].

HCA is a preliminary method to study datasets focusing on the search of natural grouping and classification of the samples. Since HCA is an unsupervised mathematical procedure, it can be used as a pattern recognition technique that can reveal the structure residing in a dataset. Samples similarities are calculated on the basis of the Euclidean distance, and the Ward hierarchical agglomerative method is used to group the samples [72].

PCA is a type of multivariate analysis that transforms a number of possibly correlated variables into a smaller number of uncorrelated variables called principal components. These principal components are built as lineal combinations of the original variables and are characterized for retaining the most important information about data variation and for non repeating information between them [73].

When chromatographic fingerprints are used for PCA, there are basically two sources of variation contained in chromatographic data. One comes from the experimental process, and the other from the samples themselves. The main variation arising from experimental process is caused by some imperfections such as baseline drifts and peak shifts that may hinder the interpretation of PCA results. For this reason data pretreatment steps, such us retention time alignment, are usually needed for the PCA to be successful [12,65,71].

The PCA scores plot is the graphic representation of samples on principal components space. The distance between samples on this plot is related to similarities or differences between them. Therefore, it is possible to identify sample groups by observing the scores plot. External information about samples, for example the region of origin of natural products, can be useful to identify the nature of groups. The PCA loadings plot is the graphic representation of the variables on principal components space, and it provides information about correlations between variables and possible relationships with sample's properties [73].

Polyophenols, as well as other low weight molecular organic acids, alcohol, esters, etc, have been found to be efficient descriptors of some climatic conditions, geographical location, agricultural, and technological features. Therefore, polyphenolic chromatographic profile is a useful platform for reliable discrimination between natural products via chemometric methods [12,60].

The work proposed by Olšovská et al. [60] is a good example of the use of polyphenolic profiles, and particularly PACs, for classifying natural products. On this work, different kind of hops are analyzed by a UHPLC-TOF MS profiling method, used for monitoring bioactive PACs in samples. The relative proportion of the various PACs in hops depends on the variety and climatic conditions, for this reason experimental data were evaluated using PCA and HCA, with the aim of determining the correlation between the type and amount of PACs and both, the hop variety and the growing locality.

Figure 10 shows PCA results obtained on this work. When the points of the graph are closer, the relative abundance of PACs in the corresponding samples is similar. In accordance with the above, it is possible to separate the four tested varieties into four segments based on the PACs profile: Saaz (left part of the graph), Sladek (right part), Premiant (upper part) and Agnus (in the middle). Figure 11 shows the results of HCA and it illustrates the hierarchical structure of the objects (hop variety and growing locality) in the cluster. Except of two samples, the hops are again clearly grouped depending on their variety. The greatest distance observed on Figure 11 is between the traditional Saaz variety and the other three tested varieties. It is interesting to observe that the same distance is observed on scores plot (Figure 10). Taking into account the mentioned results, the dependence of PACs composition on variety was observed; therefore this method may be used for prediction of qualitative marks or authenticity verification of hops.

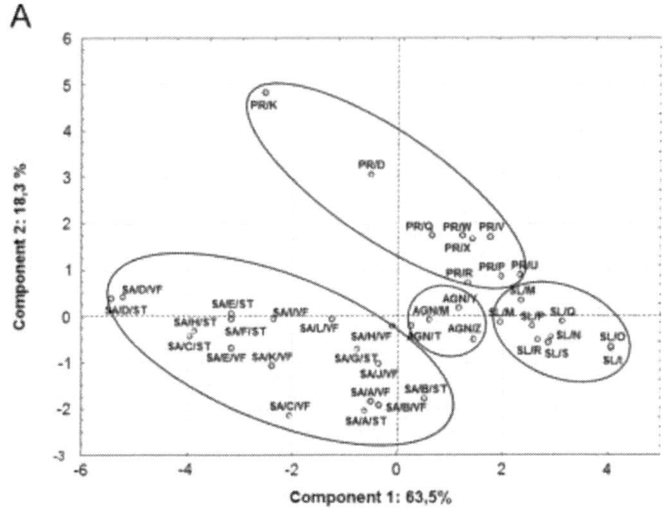

Figure 10. Principal component analysis (scores plot). Results of 2012 crop. SA-Saaz, SL – Sladek, PR – Preminant, AGN – Agnus, A,B – Z growing localities The Saaz variety: ST – standard, VF – virus free. Reprinted with permission from reference [60]. Copyright (2014) Talanta.

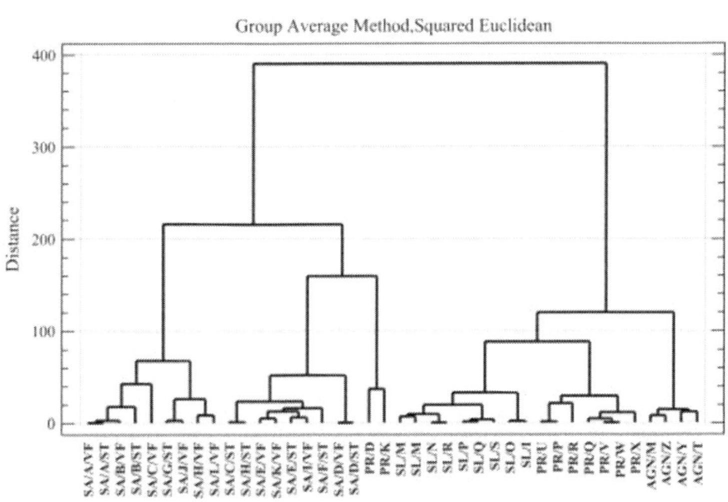

Figure 11. Cluster analysis (dendogram). Results of 2012 crop. For the legend see Figure. 10. Reprinted with permission from reference [60]. Copyright (2014) Talanta.

Tea has been used as an important drink for over 1000 years because of its benefits on human health, including anticancer, neuroprotective and antioxidant among others. These biological activities are associated in part to the antioxidant activity of chemical compounds present in teas, especially polyphenols.

For this reason, the number of scientific publications about analyzing polyphenolic compounds on tea has increased [65,71,72]. Some of these publications use chemometric tools for treating data. Zhao et al. [71] used an UPLC-DAD/MS method to analyze different types of teas (green pu-erh, green tea, white tea), and PCA was applied to classify and distinguish between tea samples. On this work three different approaches were used for PCA analysis. On the first one they used all the aligned data points of LC-DAD chromatogram. On the second one, input data was the concentration of 54 determined chemical compounds.

Finally the last one was based on using characteristic fingerprints: a chromatogram of an authentic sample of a certain group was used as a reference fingerprint (RF), the most obvious peak of the RF was selected as the reference peak (RP), the areas of all other peaks in the chromatograms were normalized against the area of the RP, and the ratios of the peaks were used for PCA analysis. The results of this study revealed that using the first approach classification of samples was not achieved; authors attributed this result to the fact that data used on this approach included all the information acquired and a lot of them were either redundant or non-critical. On the contrary, the second and the thirst approaches were capable to classify the samples according to the tea variety.

Wine is another natural beverage broadly consumed by humans. Moderate consumption of wine has been associated with several beneficial effects on human health, such as reduction of cardiovascular diseases and cancer among others.

These health promoting properties have been related to the presence of polyphenols. Furthermore, phenolic compounds contributing to the flavor and color properties of wines. For these reasons the content of wine polyphenols has been widely studied [74-76]. Some of these studies used chemometric tools for characterizing wines. The work of Vergara et al. [76] is one of these studies an also a good example of the use of PCA loadings plot. They used an HPLC-DAD-ESI-MS to study flavanols profiles of Chilean wines of Carménère and Merlot varieties. PCA was carried out to evaluate the discriminant power of the combination of several flavonols to differentiate between the two wine varieties.

The results obtained show that the samples were differentiated by the second principal component. Moreover, observing the loadings plot is possible to relate this principal component with the ratio between total myricetin and total quercetin, therefore it is possible to conclude that this variable gives the better separation between the two wine varieties studied.

Consumption of cocoa and chocolate over extended period also has been associated with a decrease in the risk of cardiovascular disease and other health benefits. These properties are related to the content of flavan-3-ols, for these reason the number of publications about flavanols on commercially available cocoa products has increased [77,78]. The work of Niemenak et al. [77] is an example of these publications that use PCA and HCA for characterizing the samples. They proposed HPLC to determine polyphenolic contents in raw cocoa beans and fermented-like seeds from Cameroon. Then PCA and HCA were used to assess the potential relationships between phenolic compounds and geographical origin, as well as to discuss the correlation between pod size, number of beans per pod and polyphenolic content.

Finally it is important to comment that this kind of works that combine the use of HPLC-MS methods for analyzing the polyphenolic content of natural samples with the use of multivariate analysis for treating data have been extended to a wide range of different kinds of natural products. Two good examples are the works of Phan-Thien et al. [79] and Navarro et al. [12]. On the first one the authors use this work strategy to differentiate between peanut samples of five different genotypes. On the second one they apply this approach to the authentication of different kind of natural extracts with different fruit origin.

SUMMARY AND FUTURE TRENDS

Among phenolic compounds in our diet, proanthocyanidins are one of the most abundant and important class and their analysis and characterization in the huge variety of food products and beverages where they are present is becoming essential not only because of the health promoting properties of these compounds but also by means of authentication of fruit-base manufacture products.

In this chapter, state-of-the-art of liquid chromatography-mass spectrometry techniques for the analysis and characterization of proanthocyanidins in natural products has been reviewed.

The analysis of proanthocyanidins is essential for a better understanding of whether naturally occurring concentrations of these compounds in food are sufficient for possible health effects. Among analytical techniques, reversed-phase liquid chromatography coupled to mass spectrometry (LC-MS(/MS)) is becoming one of the most commonly used technique for the analysis and characterization of PACs, mainly using electrospray as ionization source. However, because of the lack of adequate standards, most of the works are still focusing only on identifying and characterizing the presence of PACs in natural products, and very few of them are aiming at quantifying levels of several monomeric, dimeric, trimeric and tetrameric PACs. Further studies are still necessary to develop new strategies for the accurate quantification of PACs in natural products by overcoming the well known matrix effects when electrospray and MS techniques are used, although some approaches such as the use of echo-peak technique are described.

Electrospray ionization techniques coupled to high resolution mass spectrometry (HRMS) is becoming also a very powerful tool for the determination and characterization of PACs in natural products. The high resolution and accurate mass measurements achieved with time-of-flight and Orbitrap MS instruments are very helpful for the characterization of very complex polymeric PAC structures.

Finally, the huge amount of information of data obtained from LC-MS techniques is making necessary to treat this data by means of multivariate statistical analysis to increase the quality of the results, to facilitate its interpretation and therefore to better exploit the results. Principal component analysis and hierarchical cluster analysis are among the most frequently chemometric tools used for the characterization of PAC polyphenolic profiles in natural products. The combination of HRMS PAC polyphenolic data with chemometric tools is becoming a very interesting field to exploit in the future for the authentication of natural products.

ACKNOWLEDGMENTS

This work has been funded by the Spanish Ministry of Economy and Competitiveness under the project CTQ2012-30836, and from the Agency for Administration of University and Research Grants (Generalitat de Catalunya, Spain) under the project 2014 SGR-539.

REFERENCES

[1] Scalbert, A., Manach, C., Morand, C., Remesy, C. & Jimenez, L. (2005). Dietary polyphenols and the prevention of diseases. *Crit. Rev. Food Sci. Nutr.*, 45, 287-306.

[2] Bach-Faig, A., Berry, E. M., Lairon, D., Reguant, J., Trichopoulou, A., Dernini, S., Medina, F. X., Battino, M., Belahsen, R., Miranda, G. & Serra-Majem, L. (2011). Mediterranean Diet Foundation Expert Group. Mediterranean diet pyramid today. Science and cultural updates. *Public Health Nutr.*, 14, 2274-2284.

[3] Pastore, S., Lulli, D., Potapovich, A., Fidanza, P., Kostyuk, V., Dellambra, E., De Luca, C., Maurelli, R. & Korkina, L. (2011). Differential modulation of stress-inflammation responses by plant polyphenols in cultured normal human keratinocytes and immortalized HaCaT cells. *J. Dermatol. Sci.*, 63, 104-114.

[4] Pluemsamran, T., Onkoksoong, T. & Panich, U. (2012). Caffeic acid and ferulic acid inhibit UVA-induced matrix metalloproteinase-1 through regulation of antioxidant defense system in keratinocyte HaCaT cells. *Photochem. Photobiol.*, 88, 961-968.

[5] Pastore, S., Potapovich, A., Lulli, D., Fidanza, P., Kostyuk, V., De Luca, C., Mikhalchik, E. & Korkina, L. (2012). Plant polyphenols regulate chemokine expression and tissue repair in human keratinocytes through interaction with cytoplasmic and nuclear components of epidermal growth factor receptor (EGFR) system. *Antioxid. Redox. Signal.*, 16, 314-328.

[6] Yin, Y., Li, W., Son, Y. O., Sun, L., Lu, J., Kim, D., Wang, X., Yao, H., Wang, L., Pratheeshkumar, P., Hitron, A. J., Luo, J., Gao, N., Shi, X. & Zhang, Z. (2013). Quercitrin protects skin from UVB-induced oxidative damage. *Toxicol. Appl. Pharmacol.*, 269, 89-99.

[7] Middleton, E., Jr., Kandaswami, C. & Theoharides, T. C. (2000). The effects of plant flavonoids on mammalian cells: Implications for inflammation, heart disease, and cancer. *Pharmacol. Rev.*, 52, 673-751.

[8] U.S. Department of Agriculture and U.S. Department of Health and Human Services. *Dietary Guidelines for Americans,* 7th ed.; U.S. Government Printing Office: Washington. DC, USA, 2010.

[9] Cheynier, V. (2005). Polyphenols in foods are more complex than often thought. *Am. J. Clin. Nutr.*, 81, 223S-229S.

[10] Manach, C., Scalbert, A., Morand, C., Remesy, C. & Jimenez, L. (2004). Polyphenols: Food sources and bioavailability. *Am. J. Clin. Nutr.*, 79, 727-747.

[11] Anderson, O. & Markham, K.R. (Eds.), Flavonoids: Chemistry, Biochemistry and Applications, CRC Press, Boca Raton, 2006.

[12] Navarro, M., Núñez, O., Saurina, J., Hernández-Cassou, S. & Puignou, L. (2014). Characterization of fruit products by capillary zone electrophoresis and liquid chromatography using the compositional profiles of polyphenols: Application to authentication of natural extracts. *J. Agric. Food Chem.*, 62, 1038-1046.

[13] Porter, L. J., Ma, Z. & Chan, B. G. (1991). Cacao procyanidins: major flavanoids and identification of some minor metabolites. *Phytochemistry*, 30, 1657-1663.

[14] Guyot, S. (2012) Flavan-3-Ols and Proanthocyanidins, in Nollet, N.M.L & Toldrá, F. (eds.), Handbook of Analysis of Active Compounds in Functional Foods, CRC Press, Boca Raton, FL, USA.

[15] Gu, L., Kelm, M. A., Hammerstone, J. F., Beecher, G., Holden, J., Haytowitz, D., Gebhardt, S. & Prior, R. L. (2004). Concentrations of proanthocyanidins in common foods and estimations of normal consumption. *J. Nutr.*, 134, 613-617.

[16] Tanaka, T., Matsuo, Y. & Kouno, I. (2010). Chemistry of secondary polyphenols produced during processing of tea and selected foods. *Int. J. Mol. Sci.*, 11, 14-40.

[17] Karchesy, J. J., Hemingway, R. W., Foo, Y. L., Barofsky, E. & Barofsky, D. F. (1986). Sequencing procyanidin oligomers by fast atom bombardment mass spectrometry. *Anal. Chem.*, 58, 2563-2567.

[18] Hümmer, W. & Schreier, P. (2008). Analysis of proanthocyanidins. *Mol. Nutr. Food Res.*, 52, 1381-1398.

[19] Delcour, J. A. & Janssens de Varebeke, D. (1985). A new colorimetric assay for flavanoids in pilsner beers. *J. Inst. Brew.*, 91, 37-40.

[20] Cheynier, V., Souquet, J. M., Le Roux, E., Guyot, S. & Rigaud, J. (1999). Size separation of condensed tannins by normal-phase high-performance liquid chromatography. *Methods Enzymol.*, 299, 178-184.

[21] Gu, L., Kelm, M. A., Hammerstone, J. F., Beecher, G., Holden, J., Haytowitz, D. & Prior, R. L. (2003). Screening of Foods Containing Proanthocyanidins and Their Structural Characterization Using LC-MS/MS and Thiolytic Degradation. *J. Agric. Food Chem.*, 51, 7513-7521.

[22] Appeldoorn, M. M., Vincken, J. P., Sanders, M., Hollman, P. C. H. & Gruppen, H. (2009). Combined Normal-Phase and Reversed-Phase Liquid Chromatography/ESI-MS as a Tool To Determine the Molecular Diversity of A-type Procyanidins in Peanut Skins. *J. Agric. Food Chem.*, 57, 6007-6013.

[23] Wallace, T. C. & Giusti, M. M. (2010). Extraction and normal-phase HPLC-fluorescence-electrospray MS characterization and quantification of procyanidins in cranberry extracts. *J Food Sci*, 75, C690-C696.

[24] Kanda, T. (2001). Studies on characteristics of polyphenols in apples. *Foods Food Ingredients J. Jpn.*, 190, 15-22.

[25] Carrasco-Pancorbo, A., Cerretani, L., Bendini, A., Segura-Carretero, A., Gallina-Toschi, T. & Fernandez-Gutierrez, A. (2005). Analytical determination of polyphenols in olive oils. *J. Sep. Sci.*, 28, 837-858.

[26] Martinez-Carballo, E., Garcia-Falcon, M. S., Torrado-Agrasar, A., Pastrana-Castro, L. M. & Simal-Gandara, J. (2009). Liquid chromatography for the determination of polyphenols in beers. *Beer Health Dis. Prev.*, e281-e292.

[27] de Villiers, A., Kalili, K. M., Malan, M. & Roodman, J. (2010). Improving HPLC separation of polyphenols. *LC-GC Eur.*, 23, 466, 468, 470, 472, 474, 476, 478-

[28] Kalili, K. M. & de Villiers, A. (2011). Recent developments in the HPLC separation of phenolic compounds. *J. Sep. Sci.*, 34, 854-876.

[29] Dixon, R. A., Xie, D. Y. & Sharma, S. B. (2005). Proanthocyanidins - a final frontier in flavonoid research? *New Phytol.*, 165, 9-28.

[30] Rzeppa, S., Von Bargen, C., Bittner, K. & Humpf, H. (2011). Analysis of Flavan-3-ols and procyanidins in food samples by reversed phase high-performance liquid chromatography coupled to electrospray ionization tandem mass spectrometry (RP-HPLC-ESI-MS/MS). *J Agric Food Chem*, 59, 10594-10603.

[31] Furuuchi, R., Yokoyama, T., Watanabe, Y. & Hirayama, M. (2011). Identification and Quantification of Short Oligomeric Proanthocyanidins and Other Polyphenols in Boysenberry Seeds and Juice. *J. Agric. Food Chem.*, 59, 3738-3746.

[32] Fraser, K., Harrison, S. J., Lane, G. A., Otter, D. E., Hemar, Y., Quek, S. Y. & Rasmussen, S. (2012). HPLC-MS/MS profiling of proanthocyanidins in teas: A comparative study. *J. Food Compos. Anal.*, 26, 43-51.

[33] Delgado de la Torre, M. P., Ferreiro-Vera, C., Priego-Capote, F. & Luque de Castro, M. D. (2013). Anthocyanidins, proanthocyanidins, and

anthocyanins profiling in wine lees by solid-phase extraction-liquid chromatography coupled to electrospray ionization tandem mass spectrometry with data-dependent methods. *J. Agric. Food Chem.*, 61, 12539-12548.

[34] Hamed, A. I., Al Ayed, A. S., Moldoch, J., Piacente, S., Oleszek, W. & Stochmal, A. (2014). Profiles analysis of proanthocyanidins in the argun nut (Medemia argun-an ancient Egyptian palm) by LC-ESI-MS/MS. *J. Mass Spectrom.*, 49, 306-315.

[35] Robbins, K. S., Ma, Y., Wells, M. L., Greenspan, P. & Pegg, R. B. (2014). Separation and Characterization of Phenolic Compounds from U.S. Pecans by Liquid Chromatography-Tandem Mass Spectrometry. *J. Agric. Food Chem.*, 62, 4332-4341.

[36] Sarnoski, P. J., Johnson, J. V., Reed, K. A., Tanko, J. M. & O'Keefe, S. F. (2012). Separation and characterization of proanthocyanidins in Virginia type peanut skins by LC-MSn. *Food Chem.*, 131, 927-939.

[37] Jin, A., Ozga, J. A., Lopes-Lutz, D., Schieber, A. & Reinecke, D. M. (2012). Characterization of proanthocyanidins in pea (Pisum sativum L.), lentil (Lens culinaris L.), and faba bean (Vicia faba L.) seeds. *Food Res. Int.*, 46, 528-535.

[38] Tala, V. R. S., Candida da Silva, V., Rodrigues, C. M., Nkengfack, A. E., Campaner dos Santos, L. & Vilegas, W. (2013). Characterization of proanthocyanidins from Parkia biglobosa (jacq.) G. don. (Fabaceae) by flow injection analysis-electrospray ionization ion trap tandem mass spectrometry and liquid chromatography/electrospray ionization mass spectrometry. *Molecules*, 18, 2803-2820.

[39] Simirgiotis, M. J. & Schmeda-Hirschmann, G. (2010). Direct identification of phenolic constituents in Boldo Folium (Peumus boldus Mol.) infusions by high-performance liquid chromatography with diode array detection and electrospray ionization tandem mass spectrometry. *J. Chromatogr. A*, 1217, 443-449.

[40] Koerner, J. L., Hsu, V. L., Lee, J. & Kennedy, J. A. (2009). Determination of proanthocyanidin A2 content in phenolic polymer isolates by reversed-phase high-performance liquid chromatography. *J. Chromatogr. A*, 1216, 1403-1409.

[41] Jaiswal, R., Karakoese, H., Ruehmann, S., Goldner, K., Neumueller, M., Treutter, D. & Kuhnert, N. (2013). Identification of Phenolic Compounds in Plum Fruits (Prunus salicina L. and Prunus domestica L.) by High-Performance Liquid Chromatography/Tandem Mass

Spectrometry and Characterization of Varieties by Quantitative Phenolic Fingerprints. *J. Agric. Food Chem.*, 61, 12020-12031.

[42] Yao, X., Zhou, G. S., Tang, Y. P., Qian, Y. F., Guan, H. L., Pang, H., Zhu, S., Mo, X., Su, S. L., Jin, C., Qin, Y., Qian, D. W. & Duan, J. A. (2013). Simultaneous quantification of flavonol glycosides, terpene lactones, biflavones, proanthocyanidins, and ginkgolic acids in Ginkgo biloba leaves from fruit cultivars by ultrahigh-performance liquid chromatography coupled with triple quadrupole mass spectrometry. *BioMed Res. Int.*, 582591, 11-

[43] Engstrom, M. T., Palijarvi, M., Fryganas, C., Grabber, J. H., Mueller-Harvey, I. & Salminen, J. P. (2014). Rapid Qualitative and Quantitative Analyses of Proanthocyanidin Oligomers and Polymers by UPLC-MS/MS. *J. Agric. Food Chem.*, 62, 3390-3399.

[44] Guyot, S., Doco, T., Souquet, J. M., Moutounet, M. & Drilleau, J. F. (1996). Characterization of highly polymerized procyanidins in cider apple (Malus sylvestris var. kermerrien) skin and pulp. *Phytochemistry*, 44, 351-357.

[45] Ohnishi-Kameyama, M., Yanagida, A., Kanda, T. & Nagata, T. (1997). Identification of catechin oligomers from apple (Malus pumila cv. Fuji) in matrix-assisted laser desorption/ionization time-of-flight mass spectrometry and fast-atom bombardment mass spectrometry. *Rapid Commun. Mass Spectrom.*, 11, 31-36.

[46] Zrostlikova, J., Hajslova, J., Poustka, J. & Begany, P. (2002). Alternative calibration approaches to compensate the effect of co-extracted matrix components in liquid chromatography-electrospray ionization tandem mass spectrometry analysis of pesticide residues in plant materials. *J. Chromatogr. A*, 973, 13-26.

[47] Monagas, M., Garrido, I., Lebron-Aguilar, R., Gomez-Cordoves, M. C., Rybarczyk, A., Amarowicz, R. & Bartolome, B. (2009). Comparative Flavan-3-ol Profile and Antioxidant Capacity of Roasted Peanut, Hazelnut, and Almond Skins. *J. Agric. Food Chem.*, 57, 10590-10599.

[48] Li, H. J. & Deinzer, M. L. (2007). Tandem mass spectrometry for sequencing proanthocyanidins. *Anal. Chem.*, 79, 1739-1748.

[49] Montero, L., Herrero, M., Prodanov, M., Ibanez, E. & Cifuentes, A. (2013). Characterization of grape seed procyanidins by comprehensive two-dimensional hydrophilic interaction Ã— reversed phase liquid chromatography coupled to diode array detection and tandem mass spectrometry. *Anal Bioanal Chem*, 405, 4627-4638.

[50] Monagas, M., Quintanilla-Lopez, J. E., Gomez-Cordoves, C., Bartolome, B. & Lebron-Aguilar, R. (2010). MALDI-TOF MS analysis of plant proanthocyanidins. *J. Pharm. Biomed. Anal.*, 51, 358-372.

[51] Neto, C. C., Krueger, C. G., Lamoureaux, T. L., Kondo, M., Vaisberg, A. J., Hurta, R. A. R., Curtis, S., Matchett, M. D., Yeung, H., Sweeney, M. I. & Reed, J. D. (2005). MALDI-TOF MS characterization of proanthocyanidins from cranberry fruit (Vaccinium macrocarpon) that inhibit tumor cell growth and matrix metalloproteinase expression in vitro. *J. Sci. Food Agric.*, 86, 18-25.

[52] Jerez, M., Sineiro, J., Guitian, E. & Nunez, M. J. (2009). Identification of polymeric procyanidins from pine bark by mass spectrometry. *Rapid Commun. Mass Spectrom.*, 23, 4013-4018.

[53] Wei, S. D., Zhou, H. C., Lin, Y. M., Liao, M. M. & Chai, W. M. (2010). MALDI-TOF MS analysis of condensed tannins with potent antioxidant activity from the leaf, stem bark and root bark of Acacia confusa. *Molecules*, 15, 4369-4381.

[54] Mari, A., Eletto, D., Pizza, C., Montoro, P. & Piacente, S. (2013). Integrated mass spectrometry approach to profile proanthocyanidins occurring in food supplements: Analysis of Potentilla erecta L. rhizomes. *Food Chem.*, 141, 4171-4178.

[55] Chen, X. X., Feng, H. L., Ding, Y. M., Chai, W. M., Xiang, Z. H., Shi, Y. & Chen, Q. X. (2014). Structure characterization of proanthocyanidins from Caryota ochlandra Hance and their bioactivities. *Food Chem.*, 155, 1-8.

[56] Tourino, S., Fuguet, E., Jauregui, O., Saura-Calixto, F., Cascante, M. & Torres, J. L. (2008). High-resolution liquid chromatography/electrospray ionization time-of-flight mass spectrometry combined with liquid chromatography/electrospray ionization tandem mass spectrometry to identify polyphenols from grape antioxidant dietary fiber. *Rapid Commun. Mass Spectrom.*, 22, 3489-3500.

[57] Mouls, L., Mazauric, J. P., Sommerer, N., Fulcrand, H. & Mazerolles, G. (2011). Comprehensive study of condensed tannins by ESI mass spectrometry: average degree of polymerization and polymer distribution determination from mass spectra. *Anal. Bioanal. Chem.*, 400, 613-623.

[58] Dinelli, G., Segura Carretero, A., Di Silvestro, R., Marotti, I., Fu, S., Benedettelli, S., Ghiselli, L. & Fernandez Gutierrez, A. (2009). Determination of phenolic compounds in modern and old varieties of

durum wheat using liquid chromatography coupled with time-of-flight mass spectrometry. *J. Chromatogr. A*, 1216, 7229-7240.
[59] Iswaldi, I., Gomez-Caravaca, A. M., Arraez-Roman, D., Uberos, J., Lardon, M., Segura-Carretero, A. & Fernandez-Gutierrez, A. (2012). Characterization by high-performance liquid chromatography with diode-array detection coupled to time-of-flight mass spectrometry of the phenolic fraction in a cranberry syrup used to prevent urinary tract diseases, together with a study of its antibacterial activity. *J. Pharm. Biomed. Anal*, 58, 34-41.
[60] Olsovska, J., Kamenik, Z., Cejka, P., Jurkova, M. & Mikyska, A. (2013). Ultra-high-performance liquid chromatography profiling method for chemical screening of proanthocyanidins in Czech hops. *Talanta*, 116, 919-926.
[61] Mittal, A., Kadyan, P., Gahlaut, A. & Dabur, R. (2013). Nontargeted identification of the phenolic and other compounds of Saraca asoca by high performance liquid chromatography-positive electrospray ionization and quadrupole time-of-flight mass spectrometry. *ISRN Pharm.*, 293935, 11-
[62] Jakobek, L., Garcia-Villalba, R. & Tomas-Barberan, F. A. (2013). Polyphenolic characterisation of old local apple varieties from Southeastern European region. *J. Food Compos. Anal.*, 31, 199-211.
[63] Sun, J., Liu, X., Yang, T., Slovin, J. & Chen, P. (2014). Profiling polyphenols of two diploid strawberry (Fragaria vesca) inbred lines using UHPLC-HRMS. *Food Chem.*, 146, 289-298.
[64] Regueiro, J., Sanchez-Gonzalez, C., Vallverdu-Queralt, A., Simal-Gandara, J., Lamuela-Raventos, R. & Izquierdo-Pulido, M. (2014). Comprehensive identification of walnut polyphenols by liquid chromatography coupled to linear ion trap-Orbitrap mass spectrometry. *Food Chem.*, 152, 340-348.
[65] van der Hooft, J. J. J., Akermi, M., Uenlue, F. Y., Mihaleva, V., Roldan, V. G., Bino, R. J., de Vos, R. C. H. & Vervoort, J. (2012). Structural Annotation and Elucidation of Conjugated Phenolic Compounds in Black, Green, and White Tea Extracts. *J. Agric. Food Chem.*, 60, 8841-8850.
[66] Touriño, S., Fuguet, E., Jauregui, O., Saura-Calixto, F., Cascante, M. & Torres, J. L. (2008). High-resolution liquid chromatography/electrospray ionization time-of-flight mass spectrometry combined with liquid chromatography/electrospray ionization tandem mass spectrometry to

identify polyphenols from grape antioxidant dietary fiber. *Rapid Commun. Mass Spectrom.*, 22, 3489-3500.

[67] Makarov, A. (2000). Electrostatic Axially Harmonic Orbital Trapping: A High-Performance Technique of Mass Analysis. *Anal. Chem.*, 72, 1156-1162.

[68] Hu, Q., Noll, R. J., Li, H., Makarov, A., Hardman, M. & Cooks, R. G. (2005). The Orbitrap: A new mass spectrometer. *J. Mass Spectrom.*, 40, 430-443.

[69] Makarov, A., Denisov, E., Lange, O. & Horning, S. (2006). Dynamic Range of Mass Accuracy in LTQ Orbitrap Hybrid Mass Spectrometer. *J. Am. Soc. Mass Spectrom.*, 17, 977-982.

[70] Kalili, K. M., Vestner, J., Stander, M. A. & de Villiers, A. (2013). Toward Unraveling Grape Tannin Composition: Application of Online Hydrophilic Interaction Chromatography x Reversed-Phase Liquid Chromatography-Time-of-Flight Mass Spectrometry for Grape Seed Analysis. *Anal. Chem. (Washington, DC, U. S.)*, 85, 9107-9115.

[71] Zhao, Y., Chen, P., Lin, L., Harnly, J. M., Yu, L. & Li, Z. (2011). Tentative identification, quantitation, and principal component analysis of green pu-erh, green, and white teas using UPLC/DAD/MS. *Food Chem.*, 126, 1269-1277.

[72] Zielinski, A. A. F., Haminiuk, C. W. I., Alberti, A., Nogueira, A., Demiate, I. M. & Granato, D. (2014). A comparative study of the phenolic compounds and the in vitro antioxidant activity of different Brazilian teas using multivariate statistical techniques. *Food Res. Int.*, 60, 246-254.

[73] Massart DL, Vandeginste BGM, Deming SN, Michote Y, Kaufman L (1988) *Chemometrics: a textbook;* Elsevier: Amsterdam.

[74] Franquet-Griell, H., Checa, A., Núñez, O., Saurina, J., Hernandez-Cassou, S. & Puignou, L. (2012). Determination of Polyphenols in Spanish Wines by Capillary Zone Electrophoresis. Application to Wine Characterization by Using Chemometrics. *J. Agric. Food Chem.*, 60, 8340-8349.

[75] Gonzalez-Neves, G., Gil, G., Favre, G. & Ferrer, M. (2012). Influence of grape composition and winemaking on the anthocyanin composition of red wines of Tannat. *Int. J. Food Sci. Technol.*, 47, 900-909.

[76] Vergara, C., Von Baer, D., Mardones, C., Gutierrez, L., Hermosin-Gutierrez, I. & Castillo-Munoz, N. (2011). Flavonol profiles for varietal differentiation between Carmenere and Merlot wines produced in Chile: HPLC and chemometric analysis. *J. Chil. Chem. Soc.*, 56, 827-832.

[77] Niemenak, N., Rohsius, C., Elwers, S., Omokolo Ndoumou, D. & Lieberei, R. (2006). Comparative study of different cocoa (Theobroma cacao L.) clones in terms of their phenolics and anthocyanins contents. *J. Food Compos. Anal.*, 19, 612-619.

[78] Miller, K. B., Hurst, W. J., Flannigan, N., Ou, B., Lee, C. Y., Smith, N. & Stuart, D. A. (2009). Survey of commercially available chocolate- and cocoa-containing products in the United States. 2. Comparison of flavan-3-ol content with nonfat cocoa solids, total polyphenols, and percent cacao. *J. Agric. Food Chem.*, 57, 9169-9180.

[79] Phan-Thien, K. Y., Wright, G. C. & Lee, N. A. (2014). Peanut antioxidants: Part 2. Quantitation of free and matrix-bound phytochemicals in five selected genotypes with diverse antioxidant capacity by high performance liquid chromatography (HPLC). *LWT--Food Sci. Technol.*, 57, 312-319.

In: Proanthocyanidins
Editor: Ingrid Sullivan

ISBN: 978-1-63463-430-4

Chapter 2

PROANTHOCYANIDINS: ANTIOXIDANTS IN NEURODEGENERATION

E. González-Burgos[1*], C. Fernández-Moriano[*] and M. P. Gómez-Serranillos[†]
Department of Pharmacology, Faculty of Pharmacy,
University Complutense of Madrid, Spain

ABSTRACT

Oxidative stress results from pro-oxidant/antioxidant imbalance leading to an overproduction of reactive oxygen species (ROS) which can cause oxidative damage to cell structures and even trigger to cell death. Oxidative stress contributes to the progressive neurodegeneration associated to diseases such as Alzheimer´s disease, and Parkinson´s disease. The use of exogenous antioxidants presented in the diet that counteract oxidative stress damage by inhibiting free radical overproduction or by enhancement of the enzymatic and non-enzymatic defense mechanisms is being widely investigated as promising neuroprotector agents.

Proanthocyanidins (commonly known as condensed tannins) are secondary metabolites with oligomeric (dimers to pentamers) and polymeric (six or more units, up to 60) structures formed from the condensation of units of flavanols. These polyphenolic compounds are

---

[*] Both authors contributed equally

[†] Corresponding author

found in high proportions in cocoa products, seeds and skins of grapes, green and black tea, cinnamon and berries, among others. These compounds possess multiple therapeutic effects including anticancer, antimicrobial activity, cardioprotective role, prevention of atherosclerosis, and neuroprotective action. Many of these pharmacological benefits, including their neuroprotective action, are attributed to their significant antioxidant activity as free radical scavengers and as modulators of oxidant/antioxidant status.

The present chapter offers an overview of the health beneficial properties of proanthocyanidins in the prevention or reduction of ROS-induced oxidative damage in neurodegenerative diseases.

1. Introduction

Chemistry and Biosynthesis

Proanthocyanidins, also referred as condensed tannins, are phenolic compounds belonging to the class of flavonoids [1]. These secondary metabolites present oligomeric (dimers to pentamers) and polymeric (six or more units, up to 60) structures formed from the condensation of units of flavanols [2]. There have been described fifteen subclasses of proanthocyanidins based on the hydroxylation patterns of the aromatic rings of flavanols. The name of these subclasses refers to the resulting anthocyanidins by an acid-catalyzed depolymerization such as the conversion of (epi)catechin to cyanidin (procyanidins) and the conversion of (epi)gallocatechin to delephinidin (prodelphinidins). The most common subclasses in human foods are procyanidins (chains of the monomers epicatechin/catechin mainly found in apple and grape seeds), prodelphinidins (chains of the monomers epigallocatechin/gallocatechin presented in nuts and grapes) and propelargonidins (chains of the monomers epiafzelechin/afzelechin contained in cinnamon, berries and beans) (Figure 1) [3-5]. The flavonols units are mainly linked through a single B-type linkage (4→6 and 4→8) and a single A-type linkage (4→8 and 2→7) [6].

Proanthocyanidins are biosynthetized through the phenylpropanoid-flavonoid metabolic pathway and occur as a condensation between flavan-3-ols and flavan-3,4-diols, constituting the end product of this biosynthetic route. The models genetic organisms *Arabidopsis*, *Medicago truncatula*, maize (*Zea mays*) and grapevine (*Vitis vinifera*) have been extensively used in research to study proanthocyanidins synthesis.

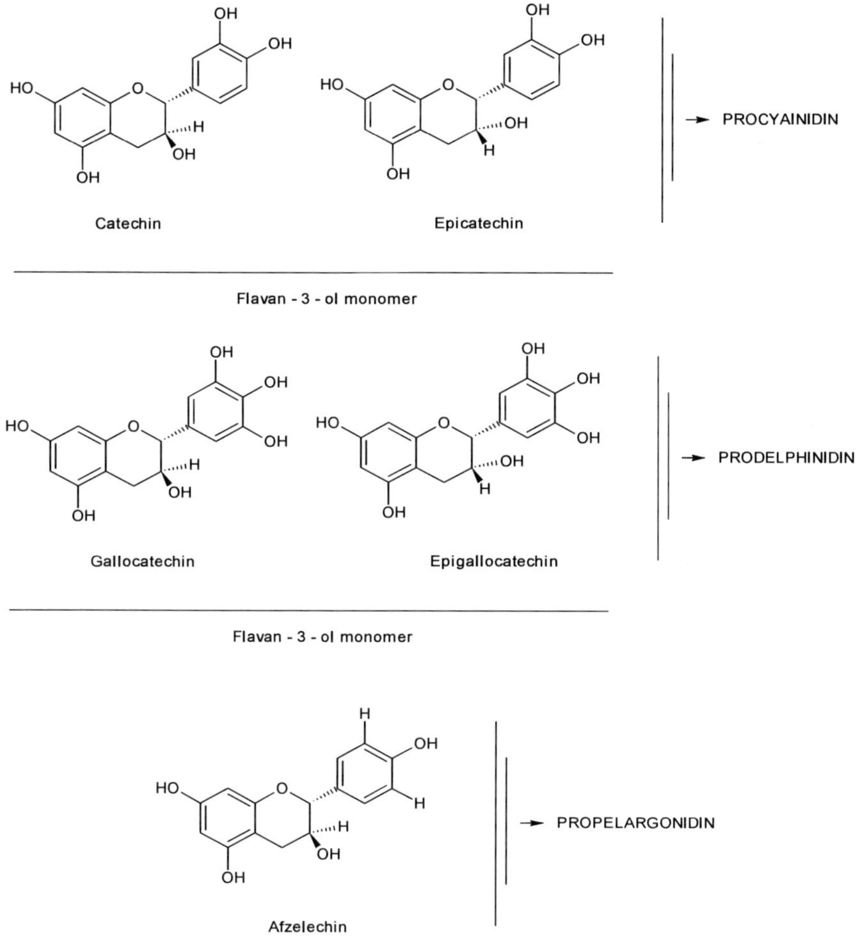

Figure 1. Basic structure of flavanols found in the different classes of proanthocyanidins.

Leucoanthocyanidins (flavan-3,4-diols including leucopelargonidin, leucocyanidin and leucodelphinidin) constitute the building blocks of proanthocyanidins together to flavan-3-ols (*cis* and *trans*) such as catechin, epicatechin and gallocatechin, among others. From leucoanthocyanidins are synthetized the 2,3-trans-flavan-3-ols (afzelechin, catechin and gallocatechin) through the action of the leucoanthocyanidin reductase (LAR) enzyme. Moreover, from leucoanthocyanidins are also synthetized the 2,3-cis-flavan-3-ols (epiafzelechin, epicatechin and epigallocatechin) by acting the enzymes anthocyanidin synthase (ANS; also named as leucoanthocyanidin dioxygenase, LDOX), and anthocyanidin reductase (ANR). Several studies have suggested that the biosynthesis of all these compounds precursors of proanthocyanidins occur on the cytosolic face of the endoplasmic reticulum (ER) surface, whereas the proanthocyanidins oligomerization/polymerization (sequentially addition of precursors units) takes place in the vacuole. There have been identified several transporters of precursors from cytosol to vacuole belonging to MATE family transporter such as the antiporter TT12 [7-9].

Sources

Proanthocyanidins are naturally occurring as oligomeric and polymeric forms in vegetables, fruits, seeds, nuts, leaves, flowers, and barks; these natural products constitute the major flavonoid type ingested in Western diets (53.6 mg/person) [10]. These secondary metabolites contribute to astringent and bitter flavor and color to foods (red, purple colors). They are found in high concentrations in apple with peels, blueberries, broccoli, cranberries, seeds and skins of grapes, green and black tea, cocoa products and pine bark [10-12]. In the Figure 2, it has been shown the chemical structure of different proanthocyanidins commonly found in human foods.

Recently, Montero et al. (2013) determined profiling and quantification of the phenolic compounds presented in different apple varieties by using comprehensive two-dimensional liquid chromatography. Among types of apples, Granny Smith apples variety contained the highest concentration in procyanidins (0.73 mg epicatechin equivalents/g dry matter) [13]. On the other hand, Huemmer et al. (2008) determined that the cultivar used and the technology of juice processing influence in both total procyanidin content and degree of polymerization as evidenced for clear and cloudy apple juices obtained from *Malus domestica* (varieties Bohnapfel and Bittenfelder) [14].

In addition to apples, other kinds of fruits contain highlight amounts of proanthocyanidins such as blueberries and cranberries. In a recent study, Yokota et al. (2013) characterized the highly polymeric proanthocyanidins content presented in the seed shells of the Japanese horse chestnut (*Aesculus turbinata* Blume) and in the fruits of blueberry (*Vaccinium angustifolium* Aiton) and cranberry (*Vaccinium macrocarpon* Aiton) by HPLC and UPLC-ESI/MS. The analysis revealed differences in both molecular size and type of proanthocyanidins. The blueberries samples had the higher molecular sizes of proanthocyanidins followed by cranberries samples, and finally by the seed shells of the Japanese horse chestnut.

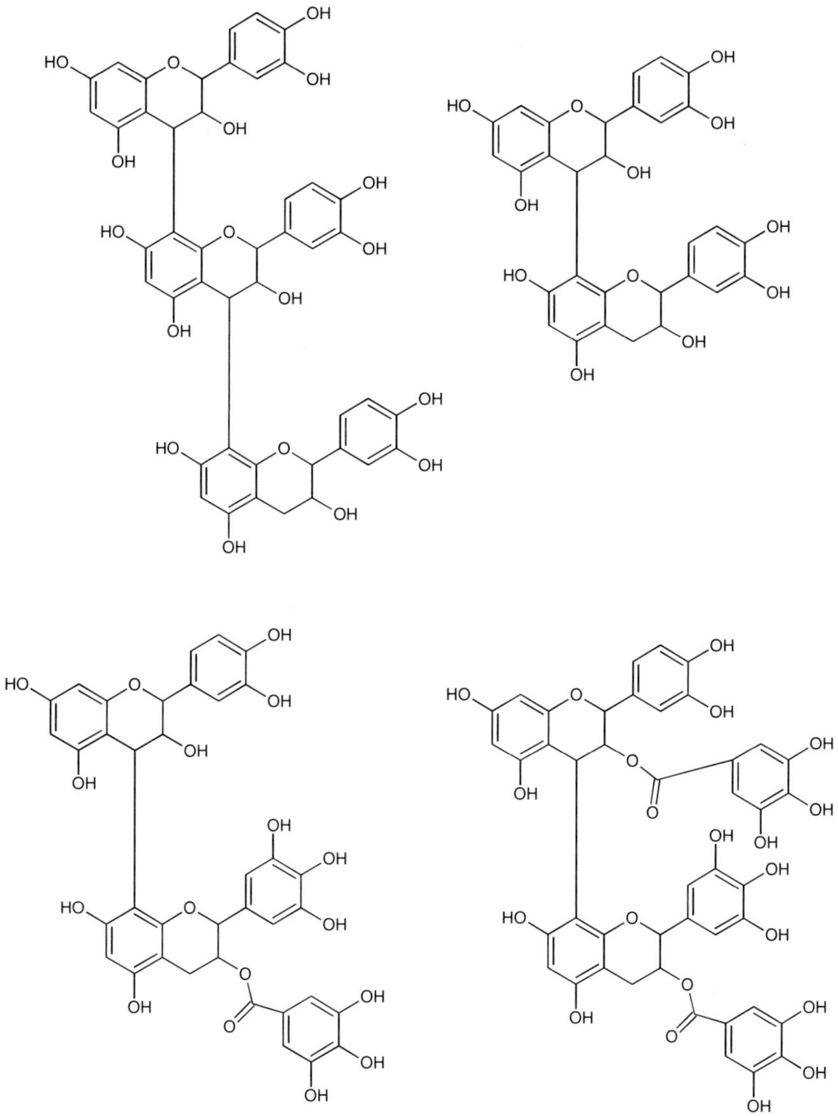

Figure 2. Chemical structure of different representative proanthocyanidins commonly found in human foods.

Moreover, both cranberry fruit and seed shells of the Japanese horse chestnut contained A-type highly polymeric proanthocyanidins (i.e., procyanidin A2, (epi)gallocatechin-(epi)catechin A-type dimer dodecylsulfide and (epi)catechin A-type trimer dodecylsulfide) as predominant compounds whereas blueberry fruit resulted to be rich in B-type highly polymeric proanthocyanidins. Furthermore, comparing seed shells of the Japanese horse and cranberry fruit chestnut samples, which are both a source of A-type highly polymeric proanthocyanidins, it has been found that the first one contained higher proportions of doubly liked A-type linkages with variations in the binding mode than the second one [6].

Tea leaves are also rich in proanthocyanidins. Green tea contains higher content in proanthocyanidins than black tea in an average proportion of 0.84% *versus* 0.5%, respectively. This is explained because during fermentation the proanthocyanidins are degraded [15]. In Chinese green tea samples (*Camelia sinensis*), Nonaka et al. (1983) identified four dimeric proanthocyanidin gallates [prodelphinidin B-2 3'-O-gallate (IV), procyanidins B-2 3,3'-di-O-gallate (V), B-2 3'-O-gallate (VI) and B-4 3'-O-gallate (VII)] as well as two dimeric flavan-3-ol gallates linked at the B-ring and named as theasinensins A (VIII) and B (IX) [16]. Moreover, Kiehne et al. (1997) determined at least 12 different proanthocyanidins such as epicatechin-3-gallate/epigallo catechin-3-gallate (EGC-EGCG), epicatechin/epicatechin/epicatechin (EC-EC-EC), epigallo catechin-3-gallate/epigallo catechin-3-gallate (EGCG-EGCG) and epigallo catechin-3-gallate/epicatechin-3-gallate (EGCG-ECG) by using high performance liquid chromatographic (HPLC) techniques [17]. Two years later, Lakenbrink et al. (1999) isolated two novel proanthocyanidins were isolated from green tea leaves, epiafzelechingallate-(4beta-->8)-epicatechingallate (EAG-4beta-->8-ECG) and epiafzelechingallate-(4beta-->6)-epicatechingallate (EAG-4beta-->6-ECG) [18].

The content of proanthocyanidins in extracts of fresh seeds of *Theobroma cacao* (cocoa) and chocolate has been also analyzed. By using HPLC/MS methods, there have been identified monomeric and oligomeric procyanidin units 2-10 in both cocoa and chocolate samples. The content in procyanidins was higher for cocoa extract than for chocolate extract [19]. On the other, the research team leader by Schmitz et al. (1999) quantified the procyanidin content in cocoa (Sanchez and Ivory Coast) and chocolates (milk, dark, and noir-type) liquors by using a modified HPLC/MS method described by Hammerstone (1999) [19]. The results of this study revealed that the average quantity of procyanidins in cocoa liquors varies depending on fermentation: the higher the degree of fermentation the lower the content in proanthocyanidins. Hence, Sanchez liquor (underfermented liquor) present higher levels in proanthocyanidins than Ivory Coast liquor (which is made from well-fermented beans). Regarding chocolates liquors, the proanthocyanidins content decreases in the order noir-type > dark > milk; both the amount of liquor component and the type of liquor used to prepare the chocolate are the factors affect its qualitative and quantitative profile of proanthocyanidins [20].

Wines are a rich source of proanthocyanidins. Comparing wines classes, differences in distribution, concentration and composition of proanthocyanidins have been identified in diverse studies. As example, Hanlin et al. (2011) analyzed the content in proanthocyanidins presented in seed, skin, and wine of Shiraz and Cabernet Sauvignon grape berries. Results revealed that the degree of polymerization was higher in skin than in seed and wine. Moreover, Shiraz grape berries samples had higher degree of polymerization than Cabernet Sauvignon grape berries samples in all seed, skin, and wine [21]. On the other hand, different types of winemaking technology have been identified to influence in the content of oligomeric and polymeric proanthocyanidins in wine. Hence, Vrhovsek et al. (2002) indicated that the higher the maceration temperature during red winemaking, the higher the content and molecular weight of proanthocyanidins. Moreover, the extraction from seeds is mainly affected by the duration of maceration whereas the extraction from skins is by the temperature of maceration [22]. In another study, Budic-Leto et al. (2003) also evaluated how maceration techniques (maceration duration and temperature) influence in oligomeric and polymeric proanthocyanins content in Croatian red wine cv. Babic; the higher the maceration duration and the higher the maceration temperature, the higher the content of proanthocyanins. Moreover, the level of these polyphenolic compound types was significantly increased in aged wines maturated for 6 and 14 months [23]. On the other hand, Sun et al. (2001) performed a study on the influence of different winemaking technologies including fermentation with carbonic maceration, fermentation with stem contact, and fermentation without stem contact on phenolic composition in Tinta Miúda red wines. Concerning procyanidins, this study revealed that the wine made by carbonic maceration contained the highest concentration in polymeric proanthocyanidins followed by the wine made by fermentation with stem contact and by fermentation without stem contact [24].

Therapeutic Applications

- Anti-obesity action: Combination of grape seed proanthocyanidins extract and docosahexaenoic-rich oil supplementation to diet-induced obese rats for 21 days reduced the harmful effects of obesity through regulation of lipid metabolism by 5'-AMP-activated protein kinase (AMPK) [25]. Moreover, grape seed proanthocyanidin extracts have demonstrated to improve dyslipidemia in rats fed on a high-fat diet and this action is related to repression of those genes controlling lipogenesis and very low density lipoprotein (VLDL) assembling in liver [26]. In another study, Jhun et al. (2013) observed that oral administration grape seed proanthocyanidin extract (GSPE) exerted an *in vivo* antiobesity effect by increasing T (Treg) cells and decreasing Th17 cells through the inhibition of the key transcriptional factors $pSTAT3^{Tyr705}$and $pSTAT3^{Ser727}$ [27].
- Antidepressant action: The authors Xu et al. (2010) demonstrated the antidepressant action of proanthocyanidins (oral administration of 25 and 50 mg/kg doses during 7 days) by using the behavioral models, the tail suspension and forced swimming tasks. This antidepressant effect was related to the ability of proanthocyanidins to inhibit monoamine oxidase activity and activate serotonergic system [28].
- Anticancer action: Proanthocyanidins have demonstrated to possess anticancer properties against a wide range of tumor types.

Grape seed proanthocyanidins resulted to be promising lung anticancer agents as evidenced by the inhibition of migration mechanisms of *in vitro* cultures of non-small cell lung cancer (NSCLC) cells exposed to cigarette smoke condensate; this inhibitory effect of cell migratory is mediated through reduction of oxidative stress via inhibition of NADPH oxidase (NOX) and through reduction in the epithelial-to-mesenchymal transition, particularly in the binding of p22phox and p47phox proteins [29]. Moreover, grape seed proanthocyanidins also exerted *in vitro* protective effects in non-small cell lung cancer (NSCLC) cells by activating apoptosis pathway [upregulation of the pro-apoptotic protein Bax, downregulation of the anti-apoptotic proteins Bcl2 and Bcl-xl, activation of the enzyme poly(ADP-ribose) polymerase(PARP) and several caspases including caspase 3 and caspase 9 and disruption of mitochondrial membrane potential] [30].

In addition to proanthocyanidins from grape seed, other natural by-products have demonstrated to be active against lung cancer such as

Cinnamomi Cortex extract and proanthocyanidin rich cranberry fraction (PAC). *In vitro* studies showed that procyanidin tetramers and pentamers contained in Cinnamomi Cortex extract suppress Nuclear factor-E2-related factor 2 (Nrf2)-regulated enzyme activity and Nrf2 expression and, they also sensitize cancer cells to the effects of doxorubicin and etoposide [31]. Moreover, the catechin trimer procyanidin C1 obtained from the Cinnamomi Cortex (CC) extract markedly inhibited TGF-β-induced epithelial-to-mesenchymal transition in lung [32]. Regarding proanthocyanidin rich cranberry fraction (PAC), it inhibited *in vitro* lung cancer by inducing apoptosis through the modulation of P21 gene expression [33].

Grape seed proanthocyanidins have also demonstrated to prevent skin solar ultraviolet (UV) radiation induced tumor formation in SKH-1 hairless mice by inhibiting oxidative damage (lipid peroxidation), regulating tissue fat content (liposynthesis and lipolysis processes), modulating MAPK and NF-kappaB signaling pathways and inhibiting inflammation mediators (i.e., myeloperoxidase, cyclooxygenase-2 (COX-2), and cyclin D1) [34-36]. Moreover, grape seed proanthocyanidins had an *in vitro* chemotherapeutic action against human epidermoid carcinoma cells by regulating Cdki-Cdk-cyclin cascade, and inducing apoptosis via loss of mitochondrial membrane potential and caspase-3 activation [37].

Several studies have also demonstrated the efficacy of proanthocyanidins against colon cancer. Oligomeric proanthocyanidins exhibited *in vitro* anticolorectal cancer activity by inducing apoptosis cell death as evidenced in an increase of BAX and caspase-3 protein expression and a decrease in Bcl-2 mRNA protein expression [38] (Kim 2005). Grape seed proanthocyanidins inhibited the growth of colon cancer cells through inhibiting vascular endothelial growth factor (VEGF) and Angiopoietin 1 (Ang1) expression and promoting apoptosis via PI 3-kinase/PKB pathway and caspase-3 activation [39-41].

Grape seed proanthocyanidins have also resulted to be a potential chemotherapeutic agent for pancreatic cancer. *In vitro* and *in vivo* studies have shown that the anti-pancreatic cancer mechanism of grape seed proanthocyanidins involve promoting mesenchymal-to-epithelial transition through inactivating the phosphorylation of ERK1/2 and NF-κB and the activation of apoptotic pathway (decrease in Bcl-2 and Bcl-xl levels and increase in Bax and caspase-3 levels) by targeting phosphatidylinositol-3-kinase (PI3K)/Akt via [42, 43].

There are also studies that have examined the effect of proanthocyanidins on prostate cancer. Hence, the procyanidin B3 resulted to be a promising therapeutic agent for prostate cancer treatment by inhibiting the activity of histone acetyltransferase and enhancing the antagonist activity of flutamide via inhibition of p300-dependent acetylation of androgen receptor [44]. The compound procyanidin has also shown anticancer effects on prostate carcinoma cells by inducing both apoptosis and necrosis pathways through a mitochondria-related mechanism [45]. The proanthocyanidins (PACs) from the American Cranberry (*Vaccinium macrocarpon*) assess anti-cancer properties and this protective effect is associated with its ability to inhibit matrix metalloproteinase-2 and matrix metalloproteinase-9 activity through a MAP kinase/ PI-3 kinase/ NF-kappaB/AP-1 signaling pathways regulation [46]. In a prospective cohort clinical study performed on a total of 43,268 men with a mean age of 70 years from United States was evidenced a possible indirect relationship between total flavonoids and proanthocyanidins with high-grade prostate cancer. Any link was observed for the consumption of these natural products classes and advanced prostate cancer [47].

- Antidiabetic potential: The oligomeric proanthocyanidin fraction obtained from persimmon fruit (*Diospyros kaki*) exerted higher antidiabetic effect than their polymeric proanthocyanidins in *in vivo* in streptozotocin (STZ)-induced diabetic model rats and db/db type 2 diabetic mice. The antidiabetic mechanism of proanthocyanidins from persimmon fruit implies attenuating oxidative stress-related inflammatory responses as evidenced in the inhibition of oxidative damage (reduction of lipid peroxidation and ROS production and increase of GSH/GSSG ratio) and the regulation of proteins related to inflammation [i.e., cyclooxygenase-2 (COX-2), nuclear factor-κB (NF-κB) and nitric oxide synthase (iNOS)) [48]. Cinnamon species (*Cinnamomum tamala* and *Cinnamomum cassia*), which are rich in A- and B-type procyanidin oligomer, respectively, have also shown *in vitro* and *in vivo* hypoglycemic activities [49, 50]. The antidiabetic potential of *Cinnamomum tamala* is related to an insulin concentration improvement in the blood and pancreas and, on the other hand, *Cinnamomum cassia* exhibited antidiabetic effects by promoting lipid accumulation in the adipose tissue and liver [50]. Moreover, grape seed proanthocyanidins possess *in vitro* and *in vivo* hypoglycemic activities associated with its insulinomimetic activity and antioxidant properties (lipid peroxidation and total nitrate/nitrite levels reduction and pancreatic GSH levels augmentation) [51, 52]. Furthermore, enzogenol, which is proanthocyanidins-rich bioflavonoid extract derived from the pine bark of New Zealand *Pinus radiata* trees, has been reported to possess hypoglycemic action in C57BL/KsJ-db/db mice (model of type 2 diabetes mellitus); enzogenol improves insulin resistance and hyperglycaemia by increasing the metabolism of glucose and lipids [53].

- Gastroprotective effect: Grape seed proanthocyanidin extracts supplementation reduced indomethacin-induced gastric injury in rats via amelioration of oxidative stress (inhibition of free radical production and antioxidant enzymatic and non-enzymatic regulation) [54, 55].
- Liver protective effect: *In vivo* and *in vitro* studies have revealed that oligomeric and polymeric proanthocyanidins protect from chronic formaldehyde, tert-butyl hydroperoxide, dimethylnitrosamine and mercuric chloride toxicity in liver tissue by increasing superoxide dismutase, catalase and glutathione peroxidase activities and reduced glutathione levels and by inhibiting malondialdehyde (MDA) formation and ROS production [56-59].

As it is evidenced, proanthocyanidins possess multiple therapeutical actions. However, in the present chapter we will focus on the health beneficial properties of proanthocyanidins in the prevention or reduction of ROS-induced oxidative damage in neurodegenerative diseases, such as Alzheimer´s disease and Parkinson´s disease.

2. Oxidative Stress and Antioxidants

Human body maintains a constant balance between oxidation and reduction processes, thus preserving the equilibrium between the production of pro-oxidants (generated as a result of cellular metabolism) and the antioxidant defense systems. The loss of this balance leads to an oxidative stress state. The oxidative stress (OS) is described as an imbalance between biochemical processes leading to the generation of free radicals and reactive oxygen species (ROS), pro-oxidant agents, and those responsible for the removal of ROS, the antioxidant cascade [60].

Free radicals can be defined as molecules or molecular fragments that contain one or more unpaired electrons in atomic or molecular orbitals that confer a high reactivity potential [61]. Reactive species (RS) are formed in the metabolism of free radicals and, although they are not all free radicals, they can oxidize molecules that can easily turn into free radicals and be very harmful for cells. There are various types of RS (derived from iron (RIS), nitrogen (RNS), etc.), but radicals derived from oxygen represent the most important class of reactive species generated in living systems. Actually, the addition of one electron to oxygen molecule during metabolic processes (mainly the respiratory chain in mitochondria) results in the formation of the superoxide anion radical ($O_2^{\bullet-}$), which is considered as the "primary ROS" and can react with other molecules and perpetuate the problem by forming secondary ROS, i.e., hydroxyl radical ($OH^{\bullet}$), peroxyl radical (ROO•), alkoxyl radical (RO•), hydrogen peroxide (H_2O_2) and peroxinitrite ($ONOO^-$) [62].

At low levels, ROS participate in physiological processes including the homeostatic control and cell growth. When oxidative stress occurs, ROS at high concentrations are important mediators of damage to cell structures, such as nucleic acids, lipids, carbohydrates and proteins; it therefore leads to a mitochondrial dysfunction, intracellular aggregates, excitotoxicity and apoptosis. For instance, hydroxyl radical ($OH^{\bullet}$) is able to react with both the purine and pyrimidine bases of DNA molecule, damaging it and producing permanent changes in genetic material that represent the first step in processes such as carcinogenesis and aging [63]. Metal-induced ROS generation is also known to lead to an attack on polyunsaturated fatty acid residues of phospholipids in plasmatic membrane; they form peroxyl radicals (ROO•) and, through a cyclisation reaction, end up in the formation of malondyaldehide (MDA), a mutagenic product as well [64]. Concerning proteins, amino acids residues present, in their side chains, sites that are susceptible to oxidation by the action of ROS and RNS (reactive nitrogen species); oxidation of cysteine residues may lead to abnormal formation of mixed disulphides between protein thiol groups and low molecular weight thiols, in particular glutathione (GSH) [65].

Regarding cell organelles, mitochondria play a crucial role in OS. As previously mentioned, the electron chain and oxidative phosphorylation that take place in these structures during ATP synthesis are proved to be related to side reactions that directly generate $O_2^{\bullet-}$; this radical dismutates to form H_2O_2 which can further react to form $OH^{\bullet}$. In addition, the enzyme monoamine oxidase catalyzes in the outer membrane the oxidative deamination of biogenic amines, thus being a large source of H_2O_2 and subsequent generation of more ROS. Mitochondria are remarkably sensitive to OS, and above all their mitochondrial DNA (mtDNA), due to the lack of histones and its location in mitochondrial matrix where ROS generation is favored. Oxidative damage in mtDNA causes deficiencies in the respiratory chain, leading to alterations in calcium homeostasis, mitochondrial membrane potential and ATP production. This mitochondrial dysfunction has been shown to be a pathological feature in many neurodegenerative diseases [66, 67].

On the other hand, actions of RS are opposed *in vivo* by intracellular antioxidant systems. In general, an antioxidant is any substance which, present in lower concentration than an oxidizable substrate, can significantly delay or reduce the oxidation of that substrate [68]. Antioxidant defense mechanisms may involve scavenging of ROS/RNS or their precursors, inhibition of ROS formation, removal of O_2, binding of metal ions that catalyze ROS production and up-regulation in endogenous antioxidant system. Intracellular antioxidants can be divided into:

- Enzymatic system: it includes several endogenous enzymes that act coordinately in order to reduce the levels of ROS and prevent oxidative damage in cells, such as superoxide dismutase (SOD), catalase (CAT), glutathione peroxidase (GPx), glutathione reductase (GR), thioredoxin and hemeoxigenase-1. For example, the superoxide dismutase in the mitochondria (Mn-SOD) and in cytoplasm (Cu/Zn-SOD) convert O_2^- to oxygen (O_2) and H_2O_2, and in turn, catalase and glutathione peroxidases convert this H_2O_2 and other peroxides in water; glutathione reductase recover the reduced glutathione from recycling the oxidized form.
- Non-enzymatic system: it comprises diverse compounds that act as radical scavengers by donating electrons to RS, reducing them and making them less reactive, and also by chelating metallic ions and avoiding the generation of more RS. Some of these compounds come from endogenous origin (i.e., glutathione, uric acid and ubiquinol) and the exogenous antioxidants include substances such as vitamins (ascorbic acid, α-tocopherol), polyphenols (e.g., flavonoids), carotenoids (e.g., lycopene, zeaxantine), and so on [69-71].

All tissues are sensitive to OS, but brain is especially susceptible. One reason is the high consumption of O_2 for ATP synthesis. Since neurons need great amounts of energy to maintain low gradients (high intracellular K^+, low Ca^{2+}...) and proper functioning, CNS processes around 20% of basal oxygen consumption; therefore, there is a higher production of ROS [72]. In addition, brain contains a high amount of unsaturated fatty acids and autooxidizable molecules (neurotransmitter such as dopamine, norepinephrine and serotonin), which are endogenous sources of ROS. These features, together with the low antioxidant capacity of brain cells, contribute to its high susceptibility to OS [73].

There is irrefutable evidence for some component of oxidative stress both in the normal process of aging and in all neurodegenerative diseases, with remarkable interest in the two more common conditions, such as Alzheimer´s and Parkinson´s diseases [74, 75]. Because OS is intimately linked to other components of degenerative process, it is difficult to determine whether OS leads to or is a consequence of these events.

Involvement of OS in Alzheimer´s disease (AD) has been clearly identified as a cause and/or consequence of pathologic lesions including senile plaques or neurofibrillary tangles. The main feature of enhanced OS in the AD brain is the increased content or reduced copper and iron capable of stimulating free radical generation, thus promoting protein and DNA oxidation, increased advance glication, higher levels of MDA and lipid peroxidation, etc. Under this condition, the amyloid-ß peptide (Aß), when released in the glutamatergic synapse, can potentially react with copper and zinc thus forming oxidized soluble aggregates and then precipitate amyloid [76]; free radical mechanisms and ROS generation have been suggested as plausible mechanisms for the neurotoxicity exerted by Aß peptides [77]. Moreover, postmortem study of brain and blood analysis of AD patients have demonstrated that neuronal degeneration in CNS of AD brain is associated with many OS markers such as oxidative damage to many biomacromolecules types. For instance, DNA and RNA oxidation is evidenced by increased levels of 8-hydroxy-2-deoxyguanosine (8-OHdG) and 8-hydroxyguanosine; high lipid peroxidation is marked by augmented levels of MDA, isoprostanes and 4-hydroxy-2,3-noneal (HNE); elevated levels of protein carbonyls and nitrotyrosine demonstrate the protein oxidation; and sugar oxidation is evidenced by increased glycation and glycoxidation [60].

With regard to PD, involvement of oxidative stress has been demonstrated in many *in vivo*, *in vitro* and *post-mortem* studies. Since degeneration of the melanin pigmented nigral neurons and depletion of dopamine in striatum are the pathological hallmarks, it has been shown that levels or ROS in the substantia nigra (SN) of PD patients is much higher than in control subjects [78]. Most of these ROS are derived from dopamine metabolism; enzymes monoamine oxidases (MAO-A and B) catalyze the oxidative deamination of dopamine (as well as levodopa, which is another autoxidisable compound used in the treatment of PD), producing higher levels of hydrogen peroxide. This H_2O_2 can be easily transformed into $OH^{\cdot}$ through Fenton reaction, which is favored by both the high levels of ferrous ion in SN of PD brain [79].

Moreover, decreases in the activity of complex I (NADH dehydrogenase) have been specifically found in mitochondria in the SN (both in neurons and glia), and the inhibition of this complex could lead to the generation of more reactive species ($O_2^{\cdot}$, $ONOO^-$, $OH^{\cdot}$); then, these RS can cause more damage in mitochondria, and it is plausible that OS and mitochondrial defects generate a vicious cycle in PD. Actually, complex I inhibition has been used as a main target when reproducing PD in research models [80]. The above mentioned biochemical abnormalities, are accompanied by other OS biomarkers in PD; for instance, it has been demonstrated the existence of increased levels of lipid peroxidation products (HNE) and DNA oxidation products (8-OHdG), together with decreased level of reduced glutathione [81].

As there are no effective drugs in the control of neurodegenerative diseases, the current and most appealing therapeutic approach to OS-related neurodegenerative diseases imply the use of antioxidants, with special value in the disease prevention. Considering that our endogenous antioxidant defenses are not always completely effective and that exposure to harmful environmental factors is increasing, it is reasonable to deem that exogenous antioxidants can result effective in weakening the adverse accumulative effects of oxidative damage in neurodegenerative disorders. With these regard, many antioxidant compounds of widely varying chemical structures (from natural origin and synthetic) have been investigated as potential therapeutic agents [82, 83]. Many preclinical works evidenced promising neuroprotective potential for some compounds addressing oxidative stress neutralization, but the therapeutic use of most of these compounds is limited due to scarce benefits in clinical trials; this fact has been attributed to pharmacokinetic limitations and low ability to effectively traverse the blood brain barrier [84, 85]. Potential neuroprotective agents should have the mandatory requisite that they can cross the BBB after systemic administration. For proanthocyanidins, there are several studies regarding the ability of these natural products to pass through the BBB. As example, Scalbert et al. (2000) demonstrated that proanthocyanidins penetrate across the BBB better than tocopherol [86].

3. Antioxidant Properties of Proanthocyanidins and Their Pharmacological Potential in Neurodegenerative Disorders

Several studies have recently demonstrated that several pharmacologically active proanthocyanidin compounds present antioxidant potential and a range of actions in the CNS, with promising effects in some culture and animal models of neurodegeneration. These effects include increased cell survival and rescue of nervous tissue from several neurotoxic damages, mainly through their antioxidant activities. To sum up, Table 1 reports the beneficial antioxidant effects of proanthocyanidins under different conditions.

Proanthocyanidins are important active principles in fruits and, for instance in grapes, they are normally found in form of procyanidins mixtures, being widely distributed in red grape seeds that mostly contain (±)-catechin and (−)-epicatechin (derived from the flavan-3-ol class of flavonoids). In fact, catechin monomers and procyanidins dimers and trimers present in red win have exerted a significant *in vitro* antioxidant activity in several lipid systems and particularly protected LDL from oxidation [87]. Grape seeds, which contain a high percentage of proanthocyanidins, are the best source for proanthocyanidins extraction, and have been used in many studies. Initial evaluations of the antioxidative activity of two grape seed extracts, containing 39% and 73% proanthocyanidins, found a strong antioxidant activity *in vitro*, even stronger than commercial natural antioxidant and monomeric flavonols. Procyanidins were effective as radical scavengers, especially of hydrophilic peroxyl radicals in aqueous systems; it was evidenced that proanthocyanidins have diverse antioxidative mechanisms, including radicals scavenging action, singlet oxygen quenching action, chelating action, and inhibitory action on polyphenol oxidase. This antioxidant potential may be responsible for their various valuable effects confirmed *in vivo* on many oxidation-related diseases [88].

The commercially available IH636 grape seed proanthocyanidin extract was found to be highly available and provide greater protection *in vitro* (on diverse cell lines including PC12 cell line) and *in vivo* (in various tissues), against free radicals and free radical-induced lipid peroxidation and DNA fragmentations, than vitamins C, E and ß-carotene [89]. It was able to provide significant protection against numerous toxin-induced tissue degeneration

Table 1. Beneficial effects of proanthocyanidins in models of neurological conditions and neurodegenerative disorders

Compound/Extract	Type of study	Parameters measured	Effects observed	Reference
Grape seed proanthocyanidin extract (IH636)	*In vitro In vivo*	$O^{2\bullet-}$ and $OH^{\bullet}$ RSA Antioxidant efficacy (laser scanning confocal microscopy study) Lipid peroxidation and DNA fragmentation	↑ Protection against H_2O_2-induced oxidative damage in cell lines ↑ Protection in brain tissue against ROS and free radical-induced lipid peroxidation and DNA damage	[89, 90]
Grape seed proanthocyanidin extract (Gravinol)	*In vivo*	Learning and memory studies Blood glucose, H_2O_2 release assay, CAT activity, lipid peroxidation and protein oxidation Structural microscopy studies of hippocampus	↑ Cognitive performance with physiological aging ↓ Brain oxidative stress	[91]
Grape seed proanthocyanidin extract (Gravinol)	*In vivo*	CAT and SOD activities, lipid peroxidation and protein oxidation Acetylcholine levels, acetylcholine esterase and choline acetyl transferase activities	↑ Up-regulation of antioxidant enzymes ↑Activity of cholinergic system	[92]
Grape seed proanthocyanidin extract	*In vivo*	TBARS level measurements, SOD activity, GPx activity and CAT activity	↓ Nitrogen mustard-induced lipid peroxidation ↑ SOD and CAT activities	[93]

Table 1. (Continued)

Compound/Extract	Type of study	Parameters measured	Observed effects	Reference
Defatted grape seed crude polyphenol extract	*In vitro*	DPPH and ABTS RSA and reducing power Cell viability (MTT assay) and intracellular ROS generation (DFCH-DA assay)	Polymeric procyanidin fraction exerted the highest antioxidant capacity ↑ Resistance of PC12 cells against oxidative stress and ↓ in ROS generation	[94]
Grape seed proanthocyanidin extract	*In vivo*	Inmunohistochemical determination of RAGE, Nf-κB, ICAM-I and VCAM-I	↑ Improvement in the pathological diabetic changes in cerebral cortex via AGEs/RAGE/Nf-κB p65 pathway	[95]
Red wine	*In vitro*	Cell viability by MTT and LDH assays	↑ Cell viability in astrocytes under $FeSO_4$ and H_2O_2-induced oxidative stress	[97]
Red wine	*In vivo*	Aromatase expression (Western blotting) and aromatase and ER transcription (RT-PCR)	↑ Aromatase expression and modulatory effect on ER expression	[98]
Pinus maritima bark extract (Pycnogenol)	*In vitro*	NO production, $O^{2\bullet-}$ and $OH^{\bullet}$ RSA, quantification of nitrite and nitrate, iNOS expression (RT-PCR) and NOS activity	↑ ROS and NO scavenging activity ↓ NO generation, NOS activity and NOS expression	[101]

Compound/Extract	Type of study	Parameters measured	Effects observed	Reference
Oligomeric proanthocyanidins from persimmon fruits	*In vivo*	Sirt1, p-NF-H, MAP2, synaptophysin, p-VEGFR-2 and VEGFR-2 expressions (Western blotting) Stereotypical behaviors, motor function and memory studies	↑ Lifespan of accelerated senescence mouse model Protective effects against age-related histological changes and memory impairment	[103]
Metasequoia glyptostroboides bark extract	*In vitro* *In vivo*	DPPH and $O^{2\bullet-}$ RSA, total antioxidant capacity (FRAP assay), lipid peroxidation (TBARS assay) and ferrous ion chelating activities SOD activity, MDA and GSH levels	↑ RSA and antioxidant activities ↑ Protection against CCl_4-induced liver injury via inhibition of oxidative stress	[104]
Lotus seedpod procyanidins	*In vivo*	Cognitive impairment (water maze test) and hippocampus histopathological evaluation Antioxidant enzymes and NOS activities, determination of ROS (DCFH-DA assay), MDA and NO levels	↑ Learning and memory abilities, protection of pyramidal neurons in hippocampus ↓ MDA, ROS and NO content and NOS activity ↑ GPx, CAT and SOD activities	[105]
(+)-Catechin from *Lorantus parasiticus* leaves extract	*In vitro*	Cell viability (MTT assay) and intracellular ROS level (DFCH-DA assay) Apoptosis detection (annexin V/PI labeling assay), MMP determination (JC-1 staining assay) and cell cycle analysis (PI staining assay)	↑ Cell viability ↓ H_2O_2-induced intracellular ROS generation, externalization of phosphatidylserine, MMP dissipation and sub-G1 apoptotic fraction	[106]

Table 1. (Continued)

Compound/Extract	Type of study	Parameters measured	Effects observed	Reference
Grape seed proanthocyanidin extract	*In vivo*	Body weight and determination of total bilirubin, ALT and AST levels Aß peptide, TNFα, IL-1ß and IFN-γ levels (ELISA)	No negative effects on body weight nor toxicity on liver function ↓ Aß levels, amyloid plaques deposition and microgliosis in an AD model	[107]
Cocoa procyanidin fraction and procyanidin B2	*In vitro*	Cell viability (MTT assay) and intracellular ROS level (DFCH-DA assay) Nuclear fragmentation (DAPI staining assay), JNK and MKK4 expression (Western blotting)	↑ Protection against HNE-induced apoptosis and ↓ ROS accumulation	[108]
Various proanthocyanidin-rich botanical extracts	*In vitro*	Cell viability (MAP2 and TH-inmunoreactions), O_2 consumption and nitrite release	Neuronal protection against rotenone-induced damage	[111]
Croton celtidifolius bark proanthocyanidin-rich fraction	*In vivo*	Behavioral tests (social recognition, activity chamber…) Mitochondrial complex-I activity, TH activity (Western blotting) and MAO activity	Amelioration of memory deficits, depressive-like behaviors and locomotor activity in MPTP-model of PD ↑ Mitochondrial complex-I activity and TH activity	[112]

models, including H_2O_2-induced oxidative stress in PC12 cell line, especially through its ability to scavenge free radicals [90]. Furthermore, it has been shown that a grape seed proanthocyanidin extract (GSPE), when daily and orally administered to female Wistar rats (adult and middle-aged) for 3 months, was able to lower brain oxidative stress. GSPE demonstrated to decrease blood glucose, hydrogen peroxide level and lipid peroxidation, and increased protein sulphydryl content in the hippocampus; it also significantly improved cognitive performance in all rats and reduced the accumulation of lipofuscin (end-product of oxidation processes) in the brain. Therefore, results suggest a neuroprotective role of the GSPE against the cognitive loss with physiological aging [91].

Similarly, a previous study had demonstrated in animal model that GSPE supplementation in moderately low quantity is effective in up-regulating the antioxidant mechanism by attenuating lipid peroxidation (decrease in MDA) and protein oxidation (decrease in protein carbonyl content). At a dose of 75 mg/kg/day, GPSE revealed an increase in SOD activity and changes in cholinergic system, evidenced especially in hippocampus by an increase in choline acetyl tranferase and a moderate decrease in acetylcholine esterase [92]. Thus, GSPE consumption from an earlier age appears to contribute to counteract the free radical-induced damage in the crucial areas of the brain at later stages of life.

Another work conducted by Tekiner et al. (2011) investigated the neuroprotective activity of a grape seed proanthocyanidin extract against the harmful cellular effects in brain exerted by nitrogen mustard, when administered to rats through dermal and respiratory routes. Biochemical analyses of cortical neural tissue showed that the GSPE was able to diminish the increase in lipid peroxidation (expressed as TBARS level) induced by nitrogen mustard. However, catalase and superoxide dismutase activities were maintained increased, thus implying that structural changes induced by mustard gas can be prevented and restored by proanthocyanidin treatment [93].

Similarly, Kim et al. (2012) evaluated a crude polyphenol extract (CPE) from defatted grape seeds, together with its individual catechin and oligomeric procyanidin fraction (Fcat + olig) and polymeric procyanidin fraction (Fpol), for antioxidant and cytoprotective effects in PC12 cells. Fpol presented the highest content of polyphenols and total flavonoids, resulting in the highest DPPH radical scavenging activity and reducing power, and both fractions had higher ABTS radical scavenging activity than CPE. On cellular substrate, pretreatment for 9 hours with Fpol (200 μg/mL) increased the resistance of

PC12 cells against oxidative stress; in addition, pretreating these cells with both fractions significantly decreased H_2O_2-induced ROS generation [94].

Moreover, neuroprotective activity has been demonstrated for a GSPE under streptozotocin-induced diabetic encephalopathy condition in rats. By using Western Blot and RT-PCR, inmunohistochemical assessments of RAGE, NF-κB, ICAM-I and VCAM-I levels were performed in pyramidal cell layers. Daily intake of 250 mg/kg of GSPE could improve degenerative changes on the cerebral cortex through down-regulating the AGEs/RAGE/Nf-κB pathway, which is up-regulated in diabetes [95]. In a similar way, grape seed proanthocyanidins displayed a protective role against diabetic peripheral neuropathy in rats, via modulation of endoplasmic reticulum stress [96].

With regard to wine, red wine possesses the greatest antioxidant capacity due to its high polyphenol content; in fact, it has been linked to neuroprotection, although mechanisms are largely unknown. In a work comparing neuroprotective activities *in vitro* of distinct varieties of Spanish red wine, Gómez-Serranillos et al. (2009) demonstrated that Merlot variety was the most active. It exerted promising activity as free radical scavenger against oxidative stress induced by $FeSO_4$, H_2O_2 and $FeSO_4$ + H_2O_2, thus increasing cell viability in cultured astrocytes. This effect was related to the high content of phenolic compounds in this variety, mainly catechins and oligomeric proanthocyanidins [97]. In another study, since oestrogens have neuroprotective properties (such as memory and learning preservation), Monteiro et al. (2008) tested the effect of red wine consumption on the expression of aromatase by western blotting and transcription of aromatase and estrogen receptors (ER) by RT-PCR in the rat hippocampus (after 8 weeks of beverage supply in male Wistar rats). Aromatase transcription was significantly increased by red wine consumption (180±9%), and ERß expression was decreased. Regarding wine components, catechin (300 μM), epicatechin (200 μM), procyanidin extract (200 mg/L) and fractioned procyanidins (FI and FII; 200 mg/L) significantly decreased aromatase activity. It was then shown that procyanidins had little contribution to the effects observed. Nevertheless, the elevation in aromatase expression and the effect on ER expression induced by red wine may explain the neuroprotective ability attributed to this beverage [98].

Proanthocyanidins are also commonly found in cranberries (*Vaccinium macrocarpon*), from which many beneficial effects on human health, including neurological health, have been reported [99]. But they are present as active compounds in other natural products apart from fruits.

Actually, a standardized extract obtained from the bark of the French maritime pine *Pinus maritima* (Pycnogenol), which is mainly composed by procyanidins and phenolic acids, has been largely studied for pharmacological potential and interesting results have been collected. An investigation of the cellular mechanisms of its therapeutic effects (such as cardiovascular benefits) demonstrated that Pycnogenol presents strong free radical scavenging activity against ROS and RNS and participates in the cellular antioxidant system by regenerating the ascorbyl radical and protecting endogenous antioxidants vitamin E and glutathione from oxidative stress [100]. It also modulates NO metabolism in LPS and IFN-γ-activated macrophages (cell line RAW 264.7). Preincubation of cells with physiological concentrations of Pycnogenol significantly decreased NO generation; this effect was due to the combination of several different biological activities, i.e., its ROS and NO scavenging activity, and the inhibition of both iNOS activity and iNOS-mRNA expression [101]. Furthermore, a comparative study on antioxidant activity and procyanidin composition was conducted on the bark extracts of two kinds of pine, *Pinus pinaster* and *Pinus radiata*. Results demonstrated that *P. radiata* bark was richer in total phenols, including procyanidins content with catechin as the main unit. For *P. pinaster*, epicatechin was the predominant extension unit, with a higher degree of polimerization. However, opposite results were encountered for the corresponding water and ethyl acetate soluble fractions, where *P. radiata* showed higher degree of polymerization. This degree of polymerization was directly related to higher antioxidant properties in terms of antiradical activity in DPPH assay [102].

Oral administration of oligomeric proanthocyanidins (with a lower level of polymerization) isolated from persimmon fruits has been shown to extend the lifespan of a murine model of accelerated senescence (senescence-accelerated mouse prone/8, SAMP8) and to display protective effects against memory impairment linked to aging; it actually improved spatial and object recognition memory in SAMP8. Moreover, axons density in the hippocampal CA1 was significantly increased by oligomers administration, as well as the phosphorylation of vascular endothelial growth factor receptor (VEGFR)-2 in the hippocampal CA3, hypothalamus, and choroid plexus. Then, memory improvement and histological changes may be directly induced in hippocampus and indirectly in hypothalamus through VEGFR-2 signaling [103]. Similarly, Chen et al. (2014) studied the free radical scavenging capacity, antioxidant activity, and proanthocyanidin composition of the bark of *Metasequoia glyptostroboides*. They analyzed a total of six extracts and fractions, obtained by water-ethanol extraction and afterwards separated by a

D101 resin column chromatography. They all exerted remarkable DPPH, $O^{2\bullet-}$ and $OH^{\bullet}$ radicals scavenging capacity, total antioxidative capacity (T-AOC), lipid peroxidation inhibitory activity, and metal ions chelating capacity. A fraction obtained with 60% ethanol extraction (MGEB), followed by further separation, possessed the highest free radical scavenging and antioxidant activities, and was directly related to the highest proanthocyanidins content. In fact, ten proanthocyanidins were isolated from MGEB, and six of them were firstly reported from this plant. Furthermore, pretreatment with MGEB significantly protected against CCl_4-induced acute liver injury in mice via inhibition of oxidative stress, including protective effect against hepatic MDA elevation and depletion of T-AOC, SOD, and GSH content [104].

Duan et al. (2013) determined the preventive effect of lotus seedpod procyanidins (LSPC) on cognitive impairment and oxidative damage induced by extremely low frequency electromagnetic field exposure (ELF-EMF) in mice. Results showed that pretreatment with 90 mg/kg of LSPC before ELF-EMF exposure significantly improved learning and memory abilities, and other biochemical parameters, when compared to ELF-EMF group. LSPC daily administration was also able to revert other negative effects of ELF-EMF; it decreased MDA, ROS, NO content and lowered NOS activity and augmented, at the same time, GPx, CAT and SOD activities in hippocampus and serum. Neuroprotective activity was also evidenced by a protection of pyramidal cells of CA1 and CA3 regions of the hippocampus, and is explained by its potential to scavenge oxygen free radicals and to stimulate antioxidant enzyme activity [105].

Wong et al. (2012) evaluated the neuroprotective activity of the leaves of *Loranthus parasiticus* (Loranthaceae) against H_2O_2-induced oxidative damage in NG108-15 hybridoma cells, in order to find evidence for the traditional use of this plant in the treatment of neurodegenerative diseases. They separated an initial ethanol extract into several fractions and assessed their neuroprotective activities; results showed that the aqueous fraction (LPAF) exerted the strongest neuroprotective effect, and an activity-guided isolation was performed on this fraction in order to identify the present active compounds. As a result, two proanthocyanidins AC trimer and (+)-catechin were isolated and exhibited potential neuroprotective activities, with cell viability increasing in a dose-dependent manner, but (+)-catechin emerged as the most promising compound. Thus, further investigations were conducted and (+)-catechin demonstrated to reverse significantly the effects of H_2O_2-induced oxidative damage on intracellular ROS generation, externalization of phosphatidylserine, mitochondrial membrane potential (MMP) dissipation and sub-G1 apoptotic fraction (scored by flow cytometry). Those actions may explain the underlying neuroprotective mechanisms of (+)-catechin and support the application of *Loranthus parasiticus* as therapeutic approach in managing oxidative stress-related neurodegenerative disorders [106].

Proanthocyanidins are being studied for their protective potential in neurodegenerative disorders. Concerning Alzheimer´s disease (AD), the effects of a proanthocyanidin-rich grape seed extract (436.6 mg catechin equivalents/g) were studied in a mice model of AD (APPSwe/PS1dE9 transgenic mice). After a long-term feeding, mice receiving the extract presented no harmful effects (on body weight and liver function) and a reduction of 44% in Aß levels in brain and serum, when compared with control AD mice. This diet also reduced the amyloid plaques and attenuated neuroinflammation (microgliosis) in the brain by 49 and 70% respectively. In conclusion, these promising effects exerted by the proanthocyanidin-rich grape seed extract seem to be interesting in delaying development of AD [107].

Similarly, Cho et al. (2009) investigated the protective effects of a cocoa procyanidin fraction (CPF) and its major procyanidin B2 against the 4-hydroxynonenal (HNE)-induced apoptosis in rat pheochromocytoma (PC12) cells; apoptosis is a crucial pathway in neuronal cell death in AD patients and HNE is reported to be elevated in the brains of AD patients and induces neuronal apoptosis in the presence of oxidative stress. They revealed that a pretreatment with both CPF and procyanidin B2 were able to markedly reduce nuclear condensation and sub-G1 fractions, and attenuate ROS accumulation; CPF and procyanidin B2 also reduced the induced polymerase cleavage, the antiapoptotic proteins (Bcl-2 and Bcl-X) and the caspase-3 activation. Moreover, they showed that CPF and procyanidin B2 bound directly to mitogen-activated protein kinase kinase 4 (MKK4) and inhibit its activity, as well as the activation of c-Jun N-terminal protein kinase (JNK), which is a protein involved in HNE-induced apoptosis in PC12 cells [108].

Wang et al. (2012) demonstrated that a grape-derived polyphenolic preparation comprised of the proanthocyanidins catechin and epicatechin in monomeric (Mo), oligomeric and polymeric forms (Po), and evaluated the neuroprotective properties of these separated proanthocyanidins metabolites on a mice model of AD (Female Tg2576 mice) after oral administration. Only Mo was able to improve cognitive function and its metabolites could reach and remain in the brain. It is remarkable that a proanthocyanidin metabolite, 3'-O-methyl-epicatechin-5-O-β-glucuronide, was found in the brain after Mo treatment and was showed to promote basal synaptic transmission and potentiation in hippocampus. This effect was developed through mechanisms related to CREB signaling (cAMP response binding protein) [109].

With regard to Parkinson´s disease (PD), epidemiological findings suggest that the consumption of berries (e.g., blueberries, strawberries…) rich in both anthocyanins and proanthocyanidins may reduce PD risk [110]; in fact, beneficial effects of proanthocyanidins in some PD models have been acknowledged. For instance, Strathearn et al. (2014) investigated the ability of different botanical extracts rich in anthocyanins, proanthocyanidins or other polyphenols to suppress the neurotoxic effects of rotenone in a primary neuron cell culture model of PD. They concluded that rotenone-induced dopaminergic cell death was suppressed by extracts prepared from blueberries, grape seed, hibiscus, blackcurrant and Chinese mulberry; data showed that extract rich in anthocyanins and proanthocyanidins displayed greater neuronal protection than extracts rich in other type of polyphenols, including phenolic acids and stilbenes, and might alleviate neurodegeneration in PD. Moreover, grape seed and blueberry extracts prevented from rotenone-induced mitochondrial impairments in dopaminergic cell line, and purple extract alleviated nitrite release from LPS-stimulated microglial cell [111]. The proanthocyanidin-rich fraction (PRF) obtained from the bark of *Croton celtidifolius* (Euphorbiaceae) has been reported to display antioxidant and antiiinflammatory activities, of interest in the treatment of PD; it may represent a promising therapeutic tool for neuroprotection in PD. Moreira et al. (2010) showed that a pretreatment with PRF in rats (5 days, 10 mg/kg) is able to avoid MPTP-induced mitochondrial complex-I inhibition as well as the decrease of tyrosine hydroxylase expression, in different parts of the brain. It even ameliorated the social memory deficits, depressive-like behaviors and locomotor activity reduction observed in animals treated only with MPTP [112].

Finally, low molecular proanthocyanidin dietary biofactor Oligonol deserves to be mentioned. It is a novel technology product (that optimized bioavailability) resulting from the oligomerization of polyphenols, mainly catechin-type monomers and oligomeric proanthocyanidins from fruits (persimmons, apples, grapes…). It has shown capacity to modulate oxidative stress, and also neuroprotection and chemoprevention potentials. For instance, mice that were under a Fenton chemistry model, when previously supplemented with Oligonol, presented a reduced extent of lipid peroxidation in brain, kidney and liver. Oligonol has been suggested to modulate behavioral deficits, decrease the inflammation scores and prolonging life-spam in SAMP8 mice (model of accelerated senescence and geriatric disorders). Moreover, it also modulated the Aβ-induced oxidative damage and prevented from mitochondrial membrane dysfunction in PC12 cells, then implying potential neuroprotection. Since safety studies indicate no adverse effects in mice, application of Oligonol in pharmaceutical preparations, cosmetics, and processed foods and beverages could be considered [113].

REFERENCES

[1] L. Bravo. *Nutr. Rev*. 56, 11 (1998).

[2] F. He, Q.H. Pan, Y. Shi, C.Q. Duan. *Molecules*. 13, 12 (2008).

[3] C. Santos-Buelga, E.M. Francia-Aricha, M.T. Escribano-Bailon. *Food Chem*. 53 (1995).

[4] L.J. Porter, L.N. Hrstich, B.G. Chan. *Phytochemistry*. 25 (1986).

[5] C. Santos-Buelga, A. Scalbert. *J. Sci. Food Agric*. 80 (2000).

[6] K. Yokota, H. Kimura, S. Ogawa, T. Akihiro. *J. Chem*. 2013 (2013).

[7] J. Zhao, Y. Pang, R.A. Dixon . *Plant Physiol*. 153, 2 (2010).

[8] Y. Liu, Z. Shi, S. Maximova, M.J. Payne, M.J. Guiltinan. *BMC Plant Biol*. 13, 202 (2013).

[9] F. He, Q.H. Pan, Y. Shi, C.Q. Duan. *Molecules*. 13, 12 (2008).

[10] L. Gu, M.A. Kelm, J.F. Hammerstone, G. Beecher, J. Holden, D. Haytowitz, S. Gebhardt, R.L. *Prior. J. Nutr*. 134, 3 (2004).

[11] G.R. Beecher . *J. Nutr*. 133, 10 (2003).

[12] J.K. Hellström, A.R. Törrönen, P.H. Mattila. *J. Agric. Food Chem*. 57, 17 (2009).

[13] L. Montero, M. Herrero, E. Ibáñez, A. Cifuentes. *J. Chromatogr. A*. 1313 (2013).

[14] W. Huemmer, H. Dietrich, F. Will, P. Schreier, E. Richling. *Biotechnol. J*. 3, 2 (2008).

[15] U.H. Engelhardt, C. Lakenbrink, O. Pokorny. Proanthocyanidins, bisflavanols, and hydrolyzable tannins in green and black teas. In: Nutraceutical Beverages. Vol. 871, Chapter 19, pp 254–264 ACS *Symposium Series* (2003).

[16] G. Nonaka, O. Kawahara, I. Nishioka. *Chem. Pharm. Bull*. 31 (1983).

[17] A. Kiehne, C. Lebensm, U.H. Engelhardt. *Z. Lebensm. Unters Forsch A*. 205 (1997).

[18] C. Lakenbrink , U.H. Engelhardt , V. Wray. *J. Agric. Food Chem*. 47, 11 (1999).

[19] J.F. Hammerstone, S.A. Lazarus, A.E. Mitchell, R. Rucker, H.H. Schmitz. *J. Agric. Food Chem*. 47, 2 (1999).

[20] G.E. Adamson, S.A. Lazarus, A.E. Mitchell, R.L. Prior, G. Cao, P.H. Jacobs, B.G. Kremers, J.F. Hammerstone, R.B. Rucker, K.A. Ritter, H.H. Schmitz. *J. Agric. Food Chem.* 47, 10 (1999).

[21] R.L. Hanlin, M.A. Kelm, K.L. Wilkinson, M.O. Downey. *J. Agric. Food Chem.* 59, 24 (2011).

[22] U. Vrhovsek, A. Vanzo, J. Nemanic. *Vitis.* 4, 1 (2002).

[23] I. Budic-Leto, T. Lovric, U. Vrhovsek. *Food Technol. Biotechnol.* 41, 4 (2003)

[24] B. Sun, I. Spranger, F. Roque-do-Vale, C. Leandro, P. Belchior. *J. Agric. Food Chem.* 49, 12 (2001).

[25] E. Casanova, L. Baselga-Escudero, A. Ribas-Latre, L. Cedó, A. Arola-Arnal, M. Pinent, C. Bladé, L. Arola, M.J. Salvadó. *J. Nutr. Biochem.* (2014).

[26] H. Quesada, J.M. del Bas, D .Pajuelo, S. Díaz, J. Fernandez-Larrea, M. Pinent, L. Arola, M.J. Salvadó, C. Bladé. *Int. J. Obes. (Lond).* 33, 9 (2009).

[27] J.Y. Jhun, S.J. Moon, B.Y. Yoon, J.K. Byun, E.K. Kim, E.J. Yang, J.H. Ju, Y.S. Hong, J.K. Min, S.H. Park, H.Y. Kim, M.L. Cho. *PLoS One.* 8, 11 (2013).

[28] Y. Xu, S. Li, R. Chen , G. Li, P.A. Barish , W. You, L .Chen, M. Lin, B. Ku, J. Pan, W.O. Ogle. *Pharmacol. Biochem. Behav.* 94, 3 (2010).

[29] M. Vaid, S.K. Katiyar . *Mol. Carcinog.* (2014). doi: 10.1002/mc.22173.

[30] T. Singh, S.D. Sharma, S.K. Katiyar. *PLoS One.* 6, 11 (2011).

[31] T. Ohnuma, T. Matsumoto, A. Itoi, A. Kawana, T. Nishiyama, K. Ogura, A. Hiratsuka. *Biochem. Biophys. Res. Commun.* 413, 4 (2011).

[32] R. Kin, S. Kato, N. Kaneto, H. Sakurai, Y. Hayakawa, F. Li, K. Tanaka, I. Saiki, S. Yokoyama. *Int. J. Oncol.* 43, 6 (2013).

[33] L.A. Kresty, A.B. Howell, M. Baird. *Molecules.* 16. 3 (2011).

[34] A. Mittal, C.A. Elmets, S.K. Katiyar. *Carcinogenesis.* 24, 8 (2003).

[35] S.D. Sharma, S.M. Meeran, S.K. Katiyar. *Mol. Cancer Ther.* 6, 3 (2007).

[36] S.D. Sharma, S.K. Katiyar. *Pharm. Res.* 27, 6 (2010).

[37] S.M. Meeran, S.K. Katiyar. *Exp. Dermatol.* 16, 5 (2007).

[38] Y.J. Kim, H.J. Park, S.H. Yoon, M.J. Kim, K.H. Leem, J.H. Chung, H.K. Kim. *World J. Gastroenterol.* 11, 30 (2005).

[39] S. Huang, N. Yang, Y. Liu, J. Gao, T. Huang, L. Hu, J. Zhao, Y. Li, C. Li, X. Zhang. *Int. J. Mol. Med.* 30, 6 (2012).

[40] H. Nomoto, M. Iigo, H. Hamada, S. Kojima, H. Tsuda. *Nutr. Cancer.* 49, 1 (2004).

[41] A.M. Engelbrecht, M. Mattheyse, B. Ellis, B. Loos, M. Thomas, R. Smith, S. Peters, C. Smith, K. Myburgh. *Cancer Lett.* 258, 1 (2007).
[42] R. Prasad, M. Vaid, S.K. Katiyar. *PLoS One*. 7, 8 (2012).
[43] R. Prasad, S.K. Katiyar. *Cancer Lett.* S0304-3835, 12 (2012).
[44] Y.C. Chung, C.C. Huang, C.H. Chen, H.C. Chiang, K.B. Chen, Y.J. Chen, C.L. Liu, L.T. Chuang, M. Liu, C.P. Hsu. *Pancreas*. 41, 3 (2012).
[45] K.C. Choi, S. Park, B.J. Lim, A.R. Seong, Y.H. Lee, M. Shiota, A. Yokomizo, S. Naito, Y. Na, H.G. Yoon. *Biochem. J.* 433, 1 (2011).
[46] X.J. Shang, G. Yao, J.P. Ge, Y. Sun, W.H. Teng, Y.F. Huang. *J. Androl.* 30, 2 (2009).
[47] B.A. Déziel, K. Patel, C. Neto, K. Gottschall-Pass, R.A. Hurta. *J. Cell Biochem.* 111, 3 (2010).
[48] Y. Wang, V.L. Stevens, R. Shah, J.J. Peterson, J.T. Dwyer, S.M. Gapstur, M.L. McCullough. *Am. J. Epidemiol.* 179, 8 (2014).
[49] T. Yokozawa , E.J. Cho, C.H. Park, J.H. Kim. *Evid. Based Complement. Alternat. Med.* 2012 (2012).
[50] Z. Lu, Q. Jia, R. Wang, X. Wu, Y. Wu, C. Huang, Y. Li. *Phytomedicine*. 18, 4 (2011).
[51] L. Chen, P. Sun, T. Wang, K. Chen, Q. Jia, H. Wang, Y. Li. *J. Agric. Food Chem.* 60, 36 (2012).
[52] M. Pinent, M. Blay, M.C. Bladé, M.J. Salvadó, L. Arola, A. Ardévol. *Endocrinology*. 145, 11 (2004).
[53] A.T. El-Alfy, A.A. Ahmed, A.J. Fatani. *Pharmacol. Res.* 52, 3 (2005).
[54] C.Y. Bang, S.Y. Choung. *J. Pharm. Pharmacol.* 66, 6 (2014).
[55] D.Y. Cheung, J.I. Kim, S.H. Park, J.K. Kim. *Gastroenterol. Res. Pract.* 2014 (2014).
[56] T.H. Kim, E.J. Jeon, D.Y. Cheung, C.W. Kim, S.S. Kim, S.H. Park, S.W. Han, M.J. Kim, Y.S. Lee, M.L. Cho, J.H. Chang, J.K. Min, J.I. Kim. *Gut Liver*. 7, 3 (2013).
[57] E. Bakar, E. Ulucam, A. Cerkezkayabekir. *Environ. Toxicol.* (2014). doi: 10.1002/tox.22010.
[58] Y. Kim, Y. Choi, H. Ham, H.S. Jeong, J. Lee. *Food Chem.* 137, 1-4 (2013).
[59] Y. Deng, Z. Xu, W. Liu, H. Yang, B. Xu, Y. Wei. *Biol. Trace Elem. Res.* 146, 2 (2012).
[60] M.O. Shin, S. Yoon, J.O. Moon. *Arch. Pharm. Res.* 33, 1 (2010).
[61] B.N. Ames, M.K. Shigenaga, T.M. Hagen. *Proc. Natl. Acad. Sci. U S A.* 68, 17 (1993).
[62] L.M. Sayre, G. Perry, M.A. Smith. *Chem. Res. Toxicol.* 21, 1 (2008).

[63] B. Halliwell, J.M.C. Gutteridge. Oxford University Press (1999).

[64] C. Dorado, C.R. Vargas, S. Rivas. *Revista Facultad de Medicina UNAM.* 46, 6 (2003).

[65] M. Valko, D. Leibfritz, J. Moncol, M.T. Cronin, M. Mazur, J. Telser. *Int. J. Biochem. Cell Biol.* 39, 1 (2007).

[66] L.J. Marnett. *Mutat. Res.* 424, 1-2 (1999).

[67] E.R. Stadtman. *Curr. Med. Chem.* 11, 9 (2004).

[68] E. Cadenas, K.J. Davies. *Free Radic. Biol. Med.* 29, 3-4 (2000).

[69] . Federico, E. Cardaioli, P. Da Pozzo, P. Formichi, G.N. Gallus, E. Radi. *J. Neurol. Sci.* 322, 1-2 (2012).

[70] B.N. Ames, M.K. Shigenaga, T.M. Hagen. *Proc. Natl. Acad. Sci. U S A.* 68, 17 (1993).

[71] B. Halliwell. *Drugs Aging.* 18, 9 (2001).

[72] B. Uttara, A.V. Singh, P. Zamboni, R.T. Mahajan. *Curr. Neuropharmacol.* 7, 1 (2009).

[73] Y. Zhao, B. Zhao. *Front. Biosci. (Elite Ed).* 4 (2012).

[74] J. Emerit, M. Edeas, F. Bricaire. *Biomed. Pharmacother.* 58, 1 (2004).

[75] E. Gonzalez-Burgos, M. P. Gomez-Serranillos. Nova Science Publisher (New York). 7 (2014).

[76] K. Dasuri, L. Zhang, J. N. Keller. *Free Radic. Biol. Med.* 62 (2013).

[77] K. Venkataraman, S. Khurana, T. C. Tai. *Int. J. Mol. Sci.* 14, 9 (2013).

[78] K. Jomova, D. Vondrakova, M. Lawson, M. Valko. *Mol. Cell Biochem.* 345, 1-2 (2010).

[79] R.A. Quintanilla, F.J. Muñoz, M.J. Metcalfe, M. Hitschfeld, G. Olivares, J.A. Godoy, N.C. Inestrosa. *J. Biol. Chem.* 280, 12 (2005).

[80] P. Jenner. *Ann. Neurol.* 53 (2003).

[81] C. Zhou, Y. Huang, S. Przedborski. *Ann. N. Y. Acad. Sci.* 1147 (2008).

[82] R. Betarbet, T.B. Sherer, G. MacKenzie, M. Garcia-Osuna, A.V. Panov, J.T. Greenamyre. *Nat. Neurosci.* 3 (2000).

[83] R.B. Mythri, C. Venkateshappa, G. Harish, A. Mahadevan, U.B. Muthane, T.C. Yasha, M.M. Srinivas Bharath, S.K. Shankar. *Neurochem. Res.* 36, 8 (2011).

[84] C. Guerra-Araiza, A.L. Álvarez-Mejía, S. Sánchez-Torres, E. Farfan-García, R. Mondragón-Lozano, R. Pinto-Almazán, H. Salgado-Ceballos. *Free Radic. Res.* 47, 6-7 (2013).

[85] P.C. Trippier, K. Jansen Labby, D.D. Hawker, J.J. Mataka, R.B. Silverman. *J. Med. Chem.* 56, 8 (2013).

[86] Y. Gilgun-Sherki, E. Melamed, D. Offen. *Neuropharmacology.* 40, 8 (2001).

[87] Y. Feng, X. Wang. *Oxid. Med. Cell Longev.* 2012 (2012).
[88] A. Scalbert, G. Williamson. *J. Nutr.* 130 (2000).
[89] P.A. da Silva Porto, J.A. Laranjinham, V.A. de Freitas. *Biochem. Pharmacol.* 66 (2003).
[90] Ariga T. *Biofactors.* 21, 1-4 (2004).
[91] D. Bagchi, M. Bagchi, S.J. Stohs, D.K. Das, S.D. Ray, C.A. Kuszynski, S.S. Joshi, H.G. Pruess. *Toxicology.* 148, 2-3 (2000).
[92] D. Bagchi, M. Bagchi, S.J. Stohs, S.D. Ray, C.K. Sen, H. G. Preuss. *Ann. N.Y. Acad. Sci.* 957, (2002).
[93] S. Asha Devi, B.K. Sagar Chandrasekar, K.R. Manjula, N. Ishii. *Exp. Gerontol.* 46, 11 (2011).
[94] A.B. Asha Devi, A.B. Jolitha, N. Ishii. *Med. Sci. Monit.* 12 (2006).
[95] A. Tekiner, D. Yucel, M.A. Bayar, O. Yucel, Y. Erdem, J. Karakaya. *Turk Neurosurg.* 21, 4 (2011).
[96] Y. Kim, Y. Choi, H. Ham, H.S. Jeong, J. Lee. *J. Med. Food.* 15, 5 (2012).
[97] M. Lu, L. Xu, B. Li, W. Zhang, C. Zhang, H. Feng, X. Cui, H. Gao. *J. Nutr. Sci. Vitaminol. (Tokyo).* 56, 2 (2010).
[98] Y. Ding, X. Dai, Z. Zhang, Y. Jiang, X. Ma, X. Cai, Y. Li. *J. Nutr. Biochem.* 25, 7 (2014).
[99] M.P. Gómez-Serranillos, S. Martín, T. Ortega, O.M. Palomino, M. Prodanov, V. Vacas, T. Hernández, I. Estrella, M.E. Carretero. *Plant Foods Hum. Nutr.* 64, 4 (2009).
[100] R. Monteiro, A. Faria, N. Mateus, C. Calhau, I. Azevedo. *J. Steroid Biochem. Mol. Biol.* 111, 1-2 (2008).
[101] E. Pappas, K. M. Schaich. *Crit. Rev. Food Sci. Nutr.* 49, 9 (2009).
[102] L. Packer, G. Rimbach, F. Virgili. *Free Radic. Biol. Med.* 27, 5-6 (1999).
[103] F. Virgili, H. Kobuchi, L. Packer. *Free Radic. Biol. Med.* 24, 7-8 (1998).
[104] M. Jerez, A. Selga, J. Sineiro, J.L. Torres, M.J. Núñez. *Food Chem.* 100, 2 (2007).
[105] T. Yokozawa, C.H. Park, J.S. Noh, S.S. Roh. *Molecules.* 22, 19 (2014).
[106] F. Chen, L. Zhang, S. Zong, S. Xu, X. Li, Y. Ye. *Evid. Based Complement. Alternat. Med.* 2014 (2014).
[107] Y. Duan, Z. Wang, H. Zhang, Y. He, R. Lu, R. Zhang, G. Sun, X. Sun. *Food & Function.* 4 (2013).
[108] D.Z. Wong, H.A. Kadir, S.K. Ling. *J. Ethnopharmacol.* 139, 1 (2012).
[109] Y.J. Wang, P. Thomas, J.H. Zhong, F.F. Bi, S. Kosaraju, A. Pollard, M. Fenech, X.F. Zhou. *Neurotox. Res.* 15, 1 (2009).

[110] E.S. Cho, Y.J. Jang, N.J. Kang, M.K. Hwang, Y.T. Kim, K.W. Lee, H.J. Lee. *Free Radic. Biol. Med.* 46, 10 (2009).

[111] J. Wang, M.G. Ferruzzi, L. Ho, J. Blount, E.M. Janle, B. Gong, Y. Pan, G.A. Gowda, D. Raftery, I. Arrieta-Cruz, V. Sharma, B. Cooper, J. Lobo, J.E. Simon, C. Zhang, A. Cheng, X. Qian, K. Ono, D.B. Teplow, C. Pavlides, R.A. Dixon, G. M. Pasinetti. *J. Neurosci.* 32, 15 (2012).

[112] X. Gao, A. Cassidy, M.A. Schwarzschild, E.B. Rimm, A. Ascherio. *Neurology.* 78, 15 (2012).

[113] K.E. Strathearn, G.G. Yousef, M.H. Grace, S.L. Roy, M.A. Tambe, M.G. Ferruzzi, Q.L. Wu, J.E. Simon, M.A. Lila, J.C. Rochet. *Brain Res.* 1555 (2014).

[114] E.L. Moreira, D. Rial, A.S. Aguiar Jr, C.P. Figueiredo, J.M. Siqueira, S. DalBó, H. Horst, J. de Oliveira, G. Mancini, T.S. dos Santos, J.G. Villarinho, F.V. Pinheiro, J. Marino-Neto, J. Ferreira, A.F. De Bem, A. Latini, M.G. Pizzolatti, R.M. Ribeiro-do-Valle, R.D. Prediger. *J. Neural. Transm.* 117, 12 (2010).

[115] O.I. Aruoma, B. Sun, H. Fujii, V.S. Neergheen, T. Bahorun, K.S. Kang, M.K. Sung. *Biofactors.* 27, 1-4 (2006).

In: Proanthocyanidins
Editor: Ingrid Sullivan
ISBN: 978-1-63463-430-4

Chapter 3

FRUIT PROCYANIDINS: MODULATING INFLAMMATION TO PROMOTE HEALTH

Sara L. Coleman[1,2,*], Roger D. Hurst[1], Gregory M. Sawyer[1] and Marlena C. Kruger[2]

[1]Food and Wellness Group, Food Innovation Portfolio, The New Zealand Institute for Plant and Food Research Ltd, Palmerston North, New Zealand

[2]Institute of Food, Nutrition and Human Health, Massey University, Palmerston North, New Zealand

ABSTRACT

Procyanidins are polyphenolic compounds found in relatively high concentrations in fruits that have come to be known as biologically active in the context of promoting human health. Epidemiological evidence suggests that people who consume diets high in fruits and vegetables are less susceptible to inflammation-related diseases and diseases of multi-factorial pathogenesis such as metabolic syndrome, atherosclerosis, asthma, and cancer. Procyanidins are one of three types of proanthocyanidins and differ from the others based on the hydroxylation pattern of their B ring. They differ amongst themselves according to the

[*] Corresponding author: Sara L. Coleman. Food and Wellness Group, Food Innovation Portfolio, The New Zealand Institute for Plant and Food Research Ltd, Private Bag 11030, Palmerston North, New Zealand. Institute of Food, Nutrition and Human Health, Massey University, Private Bag 11222, Palmerston North 4442, New Zealand. E-mail: sara.coleman@plantandfood.co.nz.

type of linkage that connects the flavan-3-ol monomeric units. Despite progress, there is still a lack of measurement technologies to delineate the structure of higher polymers accurately, although specific dimer designations have been identified. Research is working toward understanding the health-promoting properties of procyanidins, specifically moving beyond their antioxidant capabilities. Bioavailability of designated dimers *in vivo* is not well understood, though there are data to suggest that *in vitro* experiments could be relevant in the human system. Population studies have shown an inverse relationship between procyanidin consumption and ill health, especially inflammatory conditions. Here we present a focused review of the properties of individual procyanidins and suspected mechanisms by which they could potentially regulate inflammation. Modulation of cytokine secretion, mechanisms involving signal transduction pathways - NF-κB, STAT, and MAPK, and procyanidin influence on the chemotaxis of inflammatory cells into the tissues, will all be discussed.

ABBREVIATIONS

AD	Alzheimer's Disease
AP-1	activator protein-1
ARE	antioxidant response element
B2G2	procyanidin B2 3,3"-di-O-gallate
CA	catechin
COPD	chronic obstructive pulmonary disease
COX-2	cyclooxygenase-2
cAMP	cyclic adenosine monophosphate
DP	degree of polymerization
EC	epicatechin
ERK1/2	extracellular-regulated kinase 1 and 2
GMCSF	granulocyte-macrophage colony-stimulating factor
IBD	inflammatory bowel disease
IFNγ	interferon gamma
IL	interleukin
JAK	janus kinase
JNK1/2	C-Jun N-terminal kinase 1 and 2
LPS	lipopolysaccharide
MALDI-TOF MS	matrix-assisted laser desorption/ionization with time-of-flight mass spectrometer
MAPK	mitogen-activated protein kinase

NFAT	nuclear factor of activated T cells
NF-κB	nuclear factor-κB
NO	nitric oxide
Nrf2	nuclear factor E2-related factor 2
STAT	signal transducers and activation of transcription
Th1/Th2	T-helper 1 and 2
TNFα	tumor necrosis factor alpha

1. INTRODUCTION

For much of human history, secondary plant metabolites have provided inspiration for treatments to improve human health. As technologies progress and interest in nutrition has blossomed, research has moved towards understanding the biological activities of these plant-derived compounds and how their food sources can be better incorporated into the human diet, to act as functional foods maintaining or regaining optimal health.

Fruits are of particular interest because it is now well established that a diet high in fruits and vegetables is advantageous for optimal health. Fruits are an abundant source of polyphenols, to which are attributed many of the health benefits gained from a high fruit and vegetable diet. Plant polyphenol is the generic title given to any plant compound that contains multiple phenolic rings in its structure. Polyphenols can be divided into many subcategories, most of which are outside the scope of this chapter.

Flavonoids, the most abundant subcategory of polyphenols (Figure 1), encompass a number of compounds, with flavan-3-ols being of particular interest because of their relatively high concentrations in fruits. When flavan-3-ol units, e.g., (+)-catechin and (-)-epicatechin, are combined in either a heterogeneous or a homogeneous manner, they form proanthocyanidins.

Lastly, procyanidins are proanthocyanidins that specifically have two hydroxyl groups attached to their B-ring (Figure 2), as opposed to one or three, in propelargonidins and prodelphinidins respectively. Procyanidins are also referred to as condensed tannins in the literature, depending on the source discipline. Interestingly, the enzyme used by plants to polymerize flavan-3-ol units into proanthocyanidins is still to be discovered [1]. This review focuses on procyanidins as an agrarian approach for potential modulation of inflammatory events for improving health.

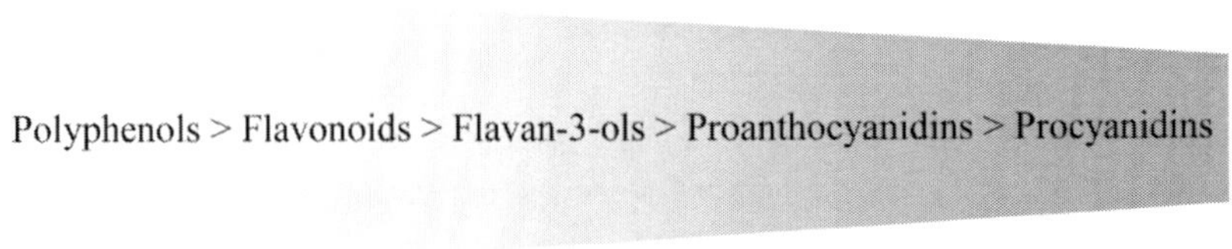

Figure 1. The nomenclature used herein arranged linearly from the broadly encompassing term, polyphenols (on the left), to the most specific term, procyanidins (on the right).

There are many challenges in designing experiments to assess the health-promoting properties of plant polyphenols, which have been reviewed [2]. However, there is much to be gained from exploring the potential of fruit procyanidins, as their possible role in mediating inflammation is now being recognized. The benefit of procyanidin intake was demonstrated in a large cohort study analyzing data from eight European countries, which showed an inverse relationship between flavan-3-ol intake and type 2 diabetes [3].

Furthermore there has been work looking at the efficacy of procyanidin-rich extracts in inhibiting the progression of Alzheimer's Disease (AD). Procyanidins from grape seed extract in combination with resveratrol and Concord grape juice extract were shown to be able to cross the blood-brain barrier and elicit an effect on a biomarker of AD progression. It was suggested by the authors of the study that while their grape seed/resveratrol/Concord grape juice extract did modulate a specific biomarker of AD progression, there is the possibility of more benefit being gained from the mixture's anti-inflammatory properties [4].

Inflammation is the process by which stressed or injured cells and tissues are repaired, dead cells cleared, and tissue function restored. This would categorize inflammation as a beneficial process to the host; however, popular opinion perceives it as negative. An acute inflammatory response that allows for immediate mitigation of a threat and then a return to basal state is ideal. During the acute inflammation response, tissue is subject to edema and increased permeability of blood vessels, causing plasma proteins to aggregate. This process results in the cardinal signs of inflammation: red, hot, swollen, tender tissue that has a loss of function [5]. These factors reflect well the etymology of the word inflammation, which comes from the Latin *inflamare*, meaning to burn [5]. This acute response is essential to host survival. However, when inflammation persists beyond the acute response, the immune balance can skew, resulting in chronic inflammation.

Chronic inflammation is an exacerbating factor in conditions such as AD, arthritis, psoriasis, eczema, irritable bowel disease (IBD), Crohn's disease, and particularly respiratory conditions such as chronic obstructive pulmonary disease (COPD) and asthma [6-8]. Chronic inflammation often lacks the outward signs of redness, heat, and pain associated with acute inflammation, but the microscopic picture in the tissue is highly abnormal. Dense masses of leukocytes, including neutrophils, eosinophils, basophils, monocytes, and lymphocytes infiltrate tissue during chronic inflammation [9].

We will overview the nomenclature and chemistry of procyanidins (important aspects when deciphering data involving procyanidins), dietary sources of procyanidins with a focus on fruit, and then provide an in-depth review of the possible mechanisms by which specific procyanidin designations may balance inflammation and thereby potentially assist in managing chronic illnesses.

2. Procyanidin Chemistry and Descriptions Used in Health Research

Procyanidins are oligomers and polymers built from monomeric flavan-3-ols (+)-catechin and (-)-epicatechin units. They vary according to the stereochemistry and hydroxylation pattern of the flavan-3-ol backbone and extension units and the bond position between units [1]. Structures of two common procyanidins can be seen in Figure 2.

Specifically to be termed a procyanidin, as opposed to other proanthocyanidins such as propelargonidins or prodelphinidins, a compound will have two (3',4') B-ring hydroxyl groups [1]. The monomeric units can be combined by either A-type or B-type linkages (Figure 2).

As research on procyanidins develops, it is imperative that scientists use the most accurate and agreed-upon descriptors, to be able to distinguish clearly which compounds are eliciting effects in a particular model of health or disease. Ideally, monomers would be listed by name, (+)-catechin or (-)-epicatechin; dimers would be listed by designations; trimers would be listed as either C1, or C2, or as a trimer with accompanying major linkage type; and polymers listed by approximate degree of polymerization (DP) and major linkage type. When technologies to delineate compound designations are unavailable, the minimum standard should be to distinguish linkage type and DP.

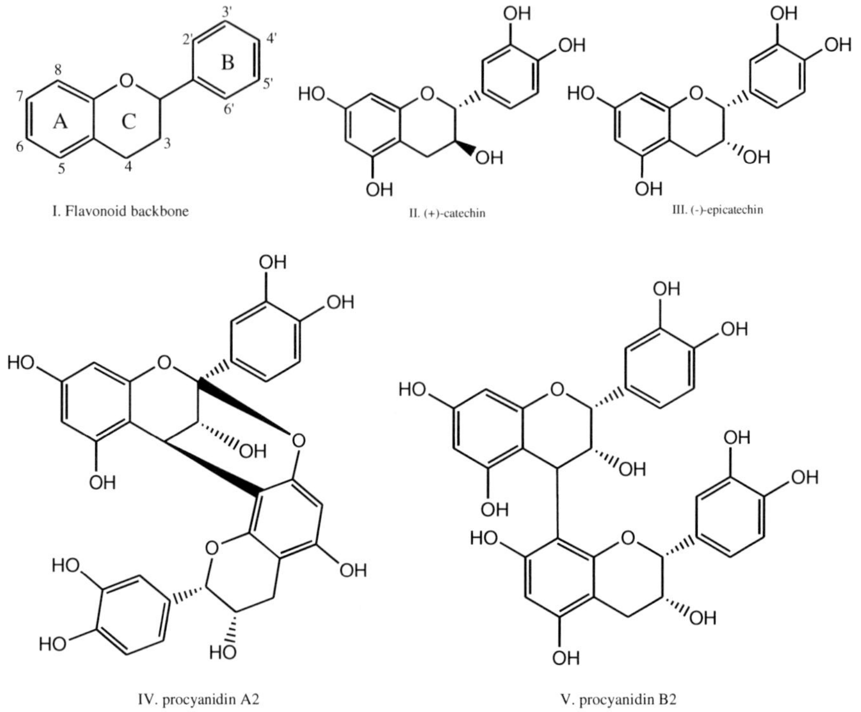

A-type link:
A1: epicatechin-(4β-8, 2β-0-7)-catechin
A2: epicatechin-(4β-8, 2β-0-7)-epicatechin

B-type link:
B1:epicatechin-(4β-8)-catechin. B2:epicatechin-(4β-8)-epicatechin
B3: catechin-(4α-8)-catechin. B4: catechin-(4α-8)-epicatechin
B5: epicatechin-(4β-6)-epicatechin. B6: catechin-(4β-6)-catechin
B7: epicatechin-(4α-6)-catechin. B8: catechin-(4α-6)-epicatechin
C1:epicatechin-(4β-8)-epicatechin-(4β-8)-epicatechin
C2: catechin-(4α-8)-catechin-(4α-8)-catechin

Figure 2. Chemical structure and designations of procyanidins. I) the backbone of flavonoid compounds, II) flavan-3-ol (+)-catechin, III) flavan-3-ol (–)-epicatechin, IV) procyanidin A2, V) procyanidin B2. A 4β-8,2β-O-7 A-type bond is illustrated in procyanidin A2. A 4β-8 B-type bond is illustrated in procyanidin B2.

Using chemistry techniques to define procyanidins involved in a study accurately was demonstrated by Feliciano and colleagues [10]. Their experiment characterized proanthocyanidins from both cranberry and apple using matrix-assisted laser desorption/ionization with a time-of-flight mass spectrometer (MALDI -TOF MS). They created proanthocyanidin treatments that were described as having one or more A-type interflavan bonds originating from cranberry, or containing only B-type interflavan bonds originating from apple [10].

The importance of such detailed description can be seen from other studies that showed the ability of isolated procyanidins B1 and B2 to form stereospecific binding directly to the p50 subunit of nuclear factor-κB (NF-κB), a predominant transcription factor that signals inflammatory events (see section 4.1) [11]. Procyanidins inhibiting NF-κB's ability to bind DNA would hinder pro-inflammatory signals from being executed. Procyanidins A1 and A2, however, were not capable of binding the p50 subunit in the same manner. In work investigating the anti-microbial properties of A-type and B-type procyanidins, A-type procyanidins at physiologically relevant doses inhibited an *Escherichia coli* bacterial strain from entering Caco2 human gut epithelial cells *in vitro*, whereas B-type proanthocyanidins were capable of such inhibition only at much higher concentrations [10]. However, both A-type and B-type procyanidins had similar hypoglycaemic activities when glucose tolerance of human liver HepG2 cells *in vitro* was measured [12]. These studies collectively show that procyanidins are extremely diverse in their abilities to influence cellular signalling and to alter cellular physiology and function. Providing the most accurate procyanidin description in research studies will lead to more robust data reporting, which will assist the success of translating *in vitro* and animal findings to the human situation and end-user utilization.

3. From Bowl to Blood Stream

Procyanidin consumption varies widely depending on both cultural food traditions and the individual [13]. Epidemiological analyses from the United States (US) and Europe have provided rough estimates of typical dietary proanthocyanidin intake. Available data are organized by generic proanthocyanidin content and not specifically by procyanidins; however, they do provide a starting point from which to determine relevant doses for scientific experimentation. Controversy surrounds the bioavailability of procyanidin compounds; they are shown to be stable during gastric transit, and monomers up to tetramers are believed to be absorbed via paracellular, passive diffusion in the small intestine, with larger polymers continuing into the large intestine [14]. However, it is currently unknown if bioavailability is a limiting step or if increased consumption of procyanidins could enhance absorption.

Work has been done to estimate the proanthocyanidin content in the human diet. The average total proanthocyanidin intake for adults in the US has been reported as 95 mg/day [15].

The major food sources of proanthocyanidins in the US are tea, legumes, and wine. Tea is the primary source of monomers and dimers, whereas legumes are the major source of polymers [15]. For this study, data were partially collected by a 24-hour food recall; the authors listed participants that did not claim to eat a specific food as non-consumers [15]; and 78% of the 8,809 people involved in the study were non-consumers of tea. Using tea as the main source of procyanidins is not a prudent choice when looking to improve human health in the US, as drinking tea is not a cultural norm as it is in many other countries, e.g., the United Kingdom or New Zealand. However, the same study also showed that only 34% of the participants were non-consumers of fruits or fruit juices, which would make fruit sources of procyanidins much more likely dietary sources, despite fruits having lower total concentrations of proanthocyanidins. Furthermore, studies looking at Finnish and Spanish diets, showed fruits (specifically apples and berries) as the major contributors to total proanthocyanidin intake [16, 17]. A large case-cohort study analyzing data from eight European countries showed results similar in nature to the US study, in that tea was the primary source of monomers, (+)-catechin and (-)-epicatechin and their gallated derivatives; whereas fruits accounted for 38% of proanthocyanidin dimers of unspecified linkage [3]. It is common for population studies to use the USDA Database for the Proanthocyanidin Content of Select Foods to estimate the proanthocyanidin content in the diet [3, 15, 17]. Unfortunately, the database is not able to give specific information on the dietary content of procyanidins by designation. If we are to imagine fruits as therapeutic tools to assist in the management of health and wellbeing, it will be necessary to identify procyanidin chemistry, specifically linkage type. Furthermore, when looking for the procyanidin content of fruits, it is pertinent to be mindful of the stage of fruit ripeness, as this has been demonstrated to relate to procyanidin concentration. Unripe white cranberries, for example, contain 184.1 mg/100 g fresh weight of total procyanidins, whereas red ripe cranberries contain 115.5 mg/100 g fresh weight of total procyanidins [18]. Proanthocyanidin content is four times higher in unripe green blackberries, and declines as they ripen to mature black fruit [19].

3.1. Procyanidins in Fruits

Identifying the concentrations of specific procyanidins in foods can be a challenge. Total proanthocyanidin content is typically listed in studies but with little information, until recently, available on designations or linkage type. Berries, apples, grapes, peaches, pears, plums, apricots, kiwifruit, mangoes and bananas are among the fruits that have been found to contain procyanidins [20]. Berries and apples have among the highest concentrations, with bananas having the least [21]. Fruits are now being investigated for specific procyanidins. Table 1 shows the concentrations of known A-type linkage dimers and B-type linkage dimers, as well as concentrations of higher polymers that are found in fruits. The data presented are assumed to apply to the whole fruit unless specified, and for some fruits (e.g., plums), procyanidin content has been evaluated in the peel and the flesh. Five cultivars of apples, 'Annurca', 'Red Delicious', Pink Lady®, 'Fuji', and 'Golden Delicious', were studied, and all but 'Red Delicious' showed higher total procyanidin content in the flesh than in the peel [18]. This is in contrast to the common notion that the majority of polyphenolic compounds are in the skins of fruits [22].

Indeed, data suggest that procyanidin content is one of the most variable characteristics between individual fruit of the same species and is influenced by the weather, sun exposure, rootstock, and cultivar [22]. For example, a study evaluating 104 varieties of apples showed a procyanidin B2 content ranging from 0.64 to 32.43 mg/100 g fresh weight [23]. For agricultural breeding to promote health, it would be advantageous to breed for plant cultivars that contain higher concentrations of procyanidins; however, even this comes with challenges for plant scientists, which have been reviewed elsewhere [24]. B-type linked procyanidins are ubiquitous and it is A-type procyanidins that are only found in a few food sources.

Most research on A-type procyanidins has occurred in cranberries, but they have also been identified in plums, and avocados [20]. Among sources of A-type procyanidins, total procyanidin content was highest in lingonberries and then cranberries, but this was mostly in the form of higher polymers. When looking at absorbable components, plums had 31.5 mg/100 g fresh weight of dimers, as opposed to 25.9 mg/100 g fresh weight of dimers in cranberries [20].

Table 1. Concentrations of Catechin (CA), Epicatechin (EC), and Procyanidins in Fruits [18, 20, 23, 28-36]

	Concentration (mg /100 g fresh weight) unless stated otherwise													
	CA	EC	B1	B2	B3	B4	B5	B6	B7	B8	C1	C2	A2	Polymers	Refererences
Apple	1.38[a] - 1.66[b]	7.70[ab]	2.71[b] - 3.40[a]	8.58[b] - 11.53[a]	0.00	0.00	0.89[a]	0.00	0.63[a]	0.00	5.80[a]	0.00	0.00	74.96[c]	20, 23, 28
Apricot	2.51	4.62	1.09	3.05	0.19	0.41	0.41	0.00	0.19	0.00	1.61	0.00	0.05	7.90	20, 28
Avocado	0.06	1.34	0.31	2.13	0.00	0.22	0.20	0.00	0.09	0.00	1.18	0.00	0.00	3.60	20, 28
Banana	0.00	0.08	0.00	0.09	0.00	0.00	0.02	0.00	0.00	0.00	0.06	0.00	0.00	2.30	20, 28
Blackberry	0.91	20.68	0.19	5.03	1.35	8.85	0.89	0.29	0.00	0.00	1.30	1.40	0.00	13.00	20, 28
Blueberry	0.77-2.10[d]	0.60	3.50	0.94[d]-1.15	0.00	0.00	0.09	0.00	0.33	0.00	0.99	0.00	0.00	163.00	20, 28, 29
Boysenberry															
Seed	5.43	7.41	-	-	3.44	13.40	-	-	-	-	-	7.41	-	11.90	30
Juice (mg/100 ml)	0.88	1.06	-	-	0.35	0.62	-	-	-	-	-	0.09	-	-	30
Cherry	6.79	7.11	3.96	3.96	0.15	0.41	0.49	0.00	0.60	0.00	3.22	0.00	0.00	8.40	20, 28
Cranberry															
American	0.43[e] - 0.49	3.33[e]-3.83	0.29 - 0.37[e]	1.56-3.23[e]	0.04	0.41	0.36 - 0.61[e]	0.00	0.09	0.00	0.37[e] - 4.00	0.00	4.61[e] - 6.90	5.97[e] A-type trimer 0.61[e] B-type trimer	18, 28, 31
European	1.38[f]	0.48[f]	0.78[f]	0.41[f]	-	-	0.07[f]	-	-	-	0.04[f]	-	0.17[f]	0.31[f] A-type trimer 0.12[f] B-type trimer	31
Currant															
Black	1.39 - 6.28[g]	0.72	0.14	0.19	0.13	0.42	0.03	0.05	0.04	0.00	0.06	0.00	0.00	-	28, 29
Red	0.77[h] - 1.37	0.29	0.27	0.06	0.24	0.09	0.01	0.03	0.03	0.00	0.03	0.10	0.00	-	28, 29
Fig	0.67	0.13	0.23	0.03	0.13	0.06	0.00	0.02	0.05	0.00	0.10	0.04	0.00	-	28
Grape	1.30[j]	0.27[j]	1.32[j]	0.11[i]	0.13[i]	0.00	0.00	0.00	0.13[j]	0.00	0.00	0.00	0.00	66.70	20, 28
Seed	-	-	-	-	-	-	-	-	-	-	-	-	-	2164.00	20

	Concentration (mg /100 g fresh weight) unless stated otherwise														
	CA	EC	B1	B2	B3	B4	B5	B6	B7	B8	C1	C2	A2	Polymers	Refererences
Lingonberry	10.10[j]	6.58[j]	10.44[j]	5.40[j]	-	-	0.93[j]	-	-	-	0.59[j]	-	5.04[j]	8.85[j] A-type trimer 2.02[j] B-type trimer	31
Litchi	0.00	1.22	0.00	1.77	0.00	0.00	0.26	0.00	0.00	0.00	0.95	0.00	0.74	-	28
Mango	1.44	0.01	0.20	0.00	0.00	0.00	0.00	0.00	0.06	0.00	0.00	0.00	0.00	7.20	20, 28
Nectarine	0.38-9.30[k]	0.03-5.30[k]	0.39-12.41[k]	0.01	0.06	0.00	0.00	0.00	0.05	0.00	0.00	0.01	0.00	16.90	20, 28, 32
Papaya	0.00	0.02	0.00	0.01	0.00	0.00	0.00	0.00	0.00	0.00	0.00	0.00	0.00	-	28
Peach	2.12 - 13.57[l]	0.10 - 6.20[l]	2.18-18.80[l]	0.07	0.23	0.00	0.02	0.00	0.21	0.00	0.00	0.05	0.00	50.60	20, 28, 32
Pear	0.58	6.14	0.50	2.75	0.04	0.08	0.42	0.00	0.16	0.00	1.89	0.00	1.14	24.20	20
Plum															
Whole Fruit	2.81[m]	4.32[m]	3.33[m]	4.74[m]	0.75[m]	6.61[m]	0.47[m]	0.00	1.91[m]	0.00	2.05[m]	0.32[m]	3.14[m]	149.10	20, 28
Flesh	10.86[n]	4.06[n]	8.65[o]-14.22[n]	7.15[o]-7.70[n]	-	8.95[o]	-	-	1.38[n]	-	-	-	-	-	32, 33
Skin	-	-	22.87[o]	16.09[o]	-	32.53[o]	-	-	-	-	-	-	-	-	32
Quince															
Fruit	0.03	2.06	0.27	3.00	0.04	0.35	0.27	0.00	0.09	0.00	2.04	0.00	0.00	-	28
Juice (mg/100 ml)	3.00[p]	45.00[p]	17.80[p]	78.90[p]	-	-	-	-	-	-	94.80[p]	-	-	1120.00[p]	34
Raspberry	0.75	8.37	0.00	0.30[q]	0.63	12.13	0.06	0.13	0.00	0.20	0.00	0.35	0.00	8.60	20, 28, 35
Rose Hips (dry weight)	2.25[r]	0.15[s]	-	2.43[r]	-	-	-	-	-	-	-	-	-	-	36
Strawberry	4.27[t]	0.14[t]	1.62[t]	0.00	3.15[t]	0.00	0.00	0.36[t]	0.33[t]	0.00	0.72[t]	3.31[t]	0.00	127.80	20, 28

Values listed a "0.00" were measured and found to be zero. A dash (-) is used for vales that were not measured in available literature. a) five cultivars of apples, b) 104 cultivars of apples, c) five cultivars of apples with and without peel, d) three cultivars of blueberries, e) six cultivars of Vaccinium macrovarpon, f) two cultivars of V. oxycoccus, g) two cultivars of blackcurrant, h) one cultivar of redcurrant, i) three samples of grapes, j) two cultivars of V. vitis-idaea (lingonberry), k) 10 cultivars of nectarines measured at both mature and ripe stages from peel and flesh, l) 10 cultivars of peaches measured at both mature and ripe stages from peel and flesh, m) five samples of plums, n) five cultivars of plums, o) five cultivars of plums measured at both mature and ripe stages, p) 12 cultivars of quinces, q) 'Muskoka' cultivated raspberry, r) five cultivars of rose hips, expressed as mg/100 g dry weight, s) two cultivars of rose hips, expressed as mg/100 g dry weight, t) three samples of strawberry.

Raspberries and strawberries are known fruits to contain propelargonidins, and blackcurrants and grapes are known fruits to contain prodelphinidins, which are the other two types of proanthocyanidins besides procyanidins. Grape seeds are one of the only sources that have been demonstrated to contain all B-type procyanidins B1-B8; however, concentrations of individual compounds have not been quantified [25].

Total proanthocyanidin contents of fruits and other foods not included here can be found from studies and reviews elsewhere [3, 15-21, 23, 26, 27].

3.2. Bioavailability and Metabolites as Bioactives

Bioavailability is defined as the proportion of compounds that are absorbed and enter the blood stream so they can be distributed and metabolized within body organs.

The term bioaccessible is used to refer to a compound being released from a food matrix during digestion [37]. The bioaccessibility of procyanidins depends on their DP, which can range from 1 up to 25+. Higher polymers are doubted ever to leave the food matrix because of entanglement with the plant cell wall polymers, and affinity for covalent attachment with other proteins. Flavon-3-ol monomers, (+)-catechin and (-)-epicatechin along with dimers up to tetramers are believed to be absorbed into the systemic circulation from the small intestine in humans, although in low concentrations [14].

Rodent experiments have demonstrated that more than 95% of consumed procyanidins remain in the intestinal lumen [38]. Procyanidins that do make it into systemic circulation are thought to undergo metabolic conjugation in the liver, producing methyl-, glucuronoyl-, and sulphoconjugates [39]. Very little work has been done *in vitro* with procyanidin metabolites, as they are not commercially available, and isolating the compounds for research can present problems such as stability. One metabolite, methyl-epicatechin glucuronide, has been shown to have protective effects over AD, working through the cyclic adenosine monophosphate (cAMP) response element binding protein [4]. If procyanidin metabolites and not the parent compounds are responsible for cellular activity, little is known of the nature of the metabolites and their modes of action.

Beyond the uncertainty of procyanidin metabolites, catechin, epicatechin, dimers and trimers have been shown to be available in plasma in their free forms at concentrations ranging of up to 8.55 μM [39]. A couple of studies have shown that catechin, epicatechin and procyanidin dimers, A1, A2, and B2 are absorbed from the small intestine without being altered [40, 41]. The bioavailability of isolated procyanidins B2 and B3 in distilled water was studied in rats, and at 1 μM plasma recovery was 89.5% and 90.8% respectively [42]. Plasma concentrations peaked at 30 min and 45 min respectively [42]. Accumulation of procyanidins was demonstrated after long-term intake, with the greatest concentrations of metabolites being found in the liver, kidney, lung, and brain [4, 43]. Furthermore, Serra and colleagues showed that both procyanidin metabolites and free-form procyanidins are bioavailable in rat plasma and that tissue accumulation varies depending on the organ [43]. These studies provide support for *in vitro* experiments showing biological activity with procyanidins, which may then be more translatable to the human situation.

The majority of procyanidins consumed bypass the small intestine and continue through the gastrointestinal tract to the large intestine. It is in the large intestine that the microbiota significantly alters parent procyanidin compounds, degrading them into smaller phenolic acids. Products generated by the microbiome have been reviewed in detail elsewhere [14]. Microbiome products can then be absorbed as these smaller phenolic acids, or excreted. It is possible that some of the health benefits seen as a result of a high procyanidin diet may result from the bioactivity of phenolic acids created by the microbiome. Human intervention studies clearly show beneficial effects from procyanidin intake, although it has been suggested that it may not be through mechanisms that require absorption into systemic circulation [14, 44] Polymeric procyanidins have recently been shown to change the physical properties of phospholipid bilayers and interact with lipid rafts in the gastrointestinal tract [13]. Hexameric procyanidins were shown to hinder a permeability insult by inhibiting extracellular-regulated kinase 1 and 2 (ERK1/2) (see section 4.4) activation in Caco-2 human intestinal epithelial cells [45].

Only a few studies have been carried out investigating procyanidin impact on lipid rafts *in vitro* and they all used specific commercially generated hexameric procyanidins; however, absorption may not be a prerequisite for cellular impact by the higher polymers found in large amounts in fruits [14]. Non absorption-mediated mechanisms such as these would side-step some of the potential issues with the systemic bioavailability of procyanidins.

4. Signal Transduction: The Dance of Inflammation and Procyanidins

Procyanidin compounds evolved in plants as a way to deal with stresses, such as drought, excess UV exposure, and/or pathogen invasion [13]. It is now clear that these compounds are also capable of modulating the human stress responses as well. As mentioned earlier, inflammation is the process by which the human body responds to injury and it involves a suite of complicated and interconnected cell-signalling cascades. Cellular signalling is an elaborate, complex network of communication mechanisms that enables cells to recognize and respond appropriately to events in their environment. When acute inflammation escalates to chronic inflammation, a variety of disease states can present with aberrant cell signalling processes, such as constitutively activated signal transducer and activators of transcription 3 (STAT3) and NF-κB in neoplastic cells (see section 4.3) [46]. Dietary procyanidin intervention could be a non-pharmaceutical approach to prevent cellular malfunction and maintain a well-controlled cell signalling process.

Initial research focused on health effects of procyanidins acting as antioxidants [47, 48]. A hawthorn extract consisting of 18% procyanidin B2, and 17% (-)-epicatechin was shown to promote binding of transcription factor nuclear factor E2-related factor 2 (Nrf2) with the antioxidant response element (ARE) in THLE-2 cells, a human liver cell line [49]. The ARE is a cis-acting element in the promoter of genes encoding major cell detoxification enzymes [50]. Nrf2 is one of the signals that link oxidative stress with inflammation and cancer [51].

Recent efforts have revealed the ability of procyanidins to modulate inflammatory signal pathways. Effects on the key signal transduction players NF-κB, Activator Protein-1 (AP-1), STAT, and mitogen-activated protein kinase (MAPK) will be discussed below.

4.1. Nuclear Factor-κB

Nuclear Factor-κB (NF-κB) is a family of transcription factors that promote the expression of over 150 important genes [52, 53]. Five members of the NF-κB family, p50, p52, p65 (RelA), RelB, and c-Rel, are predominant players in the progression of the human inflammatory response [52]. The active forms are heterodimers and of these, p50/RelA and p52/RelA are the most common. Recently Martinez-Micaelo and colleagues provided a review that covers much of what is currently known regarding procyanidins and their interactions with NF-κB, summarized as their ability to promote accumulation of p50/RelA in the cytoplasm restricting access to the nucleus, which would decrease the transcript of pro-inflammatory mediators [54]. We offer the distinction between effects of A-type and B-type dimers in reference to NF-κB here. Experiments in Jurkat T cells showed that only procyanidins B1 and B2, not A1 and A2, were able to inhibit the nuclear binding activity of NF-κB in response to an inflammatory insult [11]. The minimum energy conforms of procyanidin A1, A2, B1, and B2 were investigated for their ability to bind to the p50 component of the NF-κB heterodimer, thus restricting translocation to the nucleus. It was illustrated that the structure of the B-type linkages made it possible for hydrogen bonding with three arginine residues in the p50 subunit, while the A-type linkages did not allow for these hydrogen bonds. Procyanidins B1 and B2 are both compounds that connect via a 4β- 8 link. It could be important to model the B-type procyanidins that connect via a 4β- 6 link (e.g., procyanidins B5, B6, B7, B8) to predict if their minimum energy conforms would also be capable of inhibiting NF-κB nuclear binding activity.

4.2. Effects on Pro-Inflammatory Signalling Events

Signalling pathways are often not distinct. They interact with and influence each other, a spider's web of communication within the cell that can be very complex to unravel. Many cytokine genes are regulated cooperatively by the transcription factors AP-1 and nuclear factor of activated T-cells (NFAT), with interleukins IL-2, IL-3, IL-4, IL-5, IL-13, interferon γ (IFNγ), tumor necrosis factor α (TNFα), granulocyte-macrophage colony-stimulating factor (GMCSF), and chemokine IL-8 all having been shown to have NFAT-dependent gene regulation [55]. AP-1 is a transcription factor that is involved with the control of cellular differentiation, proliferation, and apoptosis. A studying using a procyanidin B2 derivative found in grape seed extract, procyanidin B2 3,3"-di-O-gallate (B2G2), showed an inhibition of AP-1 expression after pre-treating prostate cancer cell lines, albeit in high doses [56]. This B2 derivative also inhibited NF-κB. In a further report, the anti-inflammatory properties of isolated procyanidin A1 or apple procyanidin extracts on serum markers of the allergic response were investigated in a mouse model. While the authors did not identify their apple extract chemistry, it can be assumed that B-type linked compounds were in the majority. Mice were sensitized with interperitoneal injections of ovalbumin, and treated orally with either a control, isolated procyanidin A1, or apple extracts for 21 days.

Procyanidin A1 at 3 mg/kg/day and 10 mg/kg/day as well as apple procyanidins at 30 mg/kg/day inhibited IgE production; however, only 10 mg/kg/day A1 was able to inhibit IgG1 production [57]. Furthermore, the study looked at IFNγ and IL-4 serum concentrations. The cytokine IFNγ is secreted from T-helper 1 (Th1) cells and natural killer cells. IFNγ is needed for the maturation of undifferentiated $CD4^+$ cell into new Th1 cells and antagonizes the development of Th2 cells. The cytokine IL-4 is associated with the development of Th2 cells, and conversely inhibits the development of Th1 cells. The balance between Th1 and Th2 cell responses is considered a major factor in the progression of pathological allergic conditions, such as allergic asthma [6-8]. Procyanidin A1 at 10 mg/kg/day and apple procyanidins at 30 mg/kg/day were both able to inhibit the production of IL-4 compared with the control. Interestingly, only procyanidin A1 at 3 mg/kg/day or 10 mg/kg/day, and not the apple extracts, increased systemic IFNγ [57].

No molecular mechanisms were proposed by the authors, but it could be speculated that NFAT or AP-1 could be involved and that A-type and B-type procyanidins could interact with signal transduction elements in differing ways.

AP-1 has a reported role in differentiation of naive T cells into Th1 cells and Th2 cells [55]. Furthermore, AP-1 is suspected as a key player in the signal transduction interplay during respiratory epithelium carcinogenesis [58]. The role AP-1 has in the mechanism by which A-type procyanidins influence IFNγ has yet been tested directly.

Procyanidins were also studied in the context of IL-6, a pro-inflammatory cytokine important in the acute inflammation response. Research with THP-1 cells, a human monocytic cell line, and primary human monocytes showed the ability of procyanidin B1 and C1 to inhibit lipopolysaccharide (LPS)-induced IL-6 production after pre-treatment of procyanidin isolates for 48 hours at 17.3 μM and 11.5 μM respectively [59].

B2 showed an inhibitory effect only in primary human monocytes, not in the THP-1 cell line. Collectively these studies validate the conclusion that procyanidins' anti-inflammatory effects are widespread and will probably be specific to the individual procyanidin designation.

4.3. Influence on Signal Transducers and Activators of Transcription

Immune cell infiltration into tissues is a hallmark of chronic inflammation. Chemokines are small cytokine molecules that are secreted by various cells and tissues and promote chemotaxis of immune cells into tissue. Eotaxin-3 is one such chemokine, which promotes chemotaxis of immune eosinophils into lung tissue and is a clinical marker of asthma. There are three isoforms of eotaxin, with eotaxin-3 being the one known to remain 24 hours after an allergen challenge and hence is associated with chronic airway inflammation [60]. Work in our laboratory has shown that fruit proanthocyanidins in dimethyl sulphoxide at 2.5 μg/ml and 7.5 ug/ml are able to inhibit eotaxin-3 secretion from the human lung epithelium *in vitro* after an inflammatory insult of IL-4 [60]. The antagonistic nature of IFNγ toward the Th2 cell perpetuated process was investigated, with proanthocyanidin exposure working with IFNγ to suppress eotaxin-3 secretion. The mechanism of action of proanthocyanidin effects in alveolar epithelial cells was shown to be mediated through stimulation of STAT6 [60]. There are seven members of the STAT family in mammals: STAT1, STAT2, STAT3, STAT4, STAT5a, STAT5b, and STAT6 [61]. STATs receive an extracellular message by binding to a cellular receptor and after phosphorylation by janus kinase (JAK), dimerize and translocate to the nucleus, where they act either to activate or to repress transcription [62].

Further research using an extract from litchi fruit, a tropical fruit native to Southeast Asia, demonstrated an ability of A-type procyanidins to modulate cellular events through STAT3 in RAW 264.7 macrophages [63]. The authors showed an inhibition of the pro-inflammatory mediators nitric oxide (NO), prostaglandin E2, TNFα, IL-1β, IL-6, and cyclooxygenase-2 (COX-2) production when cells were pre-treated [63]. Investigation into the cellular signalling pathway demonstrated the involvement of NF-κB, and STAT3, but not STAT1 [63]. The lowered incidence of phosphorylated STAT3 was through inhibited activation of JAK2, which is the molecular species that interacts with the cellular receptor [63]. Another study, while not investigating isolated procyanidin compounds, used a non-fruit source of procyanidins with 9.7 mg/g procyanidin B1, and 84.3 mg/g procyanidin B2 (concentrations that could be achieved through a 100-mL serving of quince juice, Table 1) to investigate procyanidin effects on pro-inflammatory mediators in the context of IBD modelled in mice. The oral procyanidin treatment reduced leukocyte infiltration into the colon, abrogated NO production, and inhibited COX-2 secretion [64]. The molecular mechanisms responsible were shown to be a reduction in phosphorylated STAT1 and STAT3, but not in NF-κB [64]. Peritoneal macrophages were collected from the mouse model of IBD and treated *ex vivo* with the same B-type procyanidin extracts for 1 h. Experiments showed a significant decrease in the production of IL-6, IL-1β, and TNFα, but not of IFNγ [64]. It has been suggested that a balance between Th1 and Th2 responses is the optimal scenario for managing inappropriate inflammation and its associated disease [65]. Procyanidins may help to re-balance skewed Th2 responses and utilize the antagonistic powers of Th1 mechanisms. With respect to cancer development, procyanindin-mediated systemic inhibition of both STAT3 and NF-κB could result in increased damage to normal cells and tissue [64]. Simultaneous inhibition of all major inflammatory mechanisms is undesirable. Procyanidins could act as mediators of health through selective-partial inhibition of inflammatory pathways – an effect that can vary from tissue to tissue - to balance and bring back appropriate inflammatory responses and to mitigate the development of cancer.

4.4. Mitogen-Activated Protein Kinase Pathways

The MAP kinases are a conserved family of enzymes that convert extracellular signals into intracellular responses through serial phosphorylation cascades, with immune function and stress responses being only two of the many cellular processes they control [66]. Three distinct pathways exist: C-Jun N-terminal kinase 1 and 2 (JNK 1/2), the p38 pathway, and extracellular-regulated kinase 1 and 2 (ERK 1/2). The A-type procyanidin litchi fruit extract that inhibited NF-κB, and STAT3 in RAW 264.7 macrophages, was also able to inhibit pro-inflammatory mediator production by inhibiting the phosphorylation of ERK1/2, but not of p38 or JNK [63]. Moreover, procyanidin's effect on ERK1/2 phosphorylation in THP-1 human monocytes has been investigated [67]. In these studies, THP-1 cells were pre-treated with grape seed procyanidin extracts or isolated compounds for 4 h and then insulted with 1 μg/ml LPS. Both grape seed procyanidin extracts and isolated procyanidin B1 were able to inhibit LPS-induced phosphorylation of ERK 1/2.

Procyanidins B2 and C1 were also able to inhibit LPS-induced phosphorylation of ERK 1/2 but not to the same degree as the grape seed procyanidin extracts or procyanidin B1. Procyanidin B2- and C1-treated cells expressed phosphorylated ERK 1/2 at c. 10% higher than the control and 60% lower than LPS-only treated cells. While these results are promising, it should be noted that these experiments used 10 μg/mL of isolated procyanidin compounds, which would correspond to only a supplemental dose. Furthermore, a whole grape seed extract is a known source of both A-type and B-type linked procyanidins [68]. It would have been interesting to discern the results from an isolated A-type linked procyanidin such as A2. The relevance of procyanidins being capable of controlling the phosphorylation on ERK 1/2 cannot be overstated. Activated ERK 1/2 is capable of phosphorylating over 80 known substrates in the cell cytoplasm and the nucleus, affecting a variety of cellular processes such as cell survival, differentiation, metabolism, and transcription [69, 70].

Conclusion

Many chronic diseases that burden our society today are perpetuated by chronic inflammation. Fruit procyanidins have great potential in the appropriate modulation of inflammation to maintain and return health. There is evidence for procyanidins attenuating three major inflammatory pathways, NF-κB, MAPK, and STAT, with possible collaboration from AP-1. Research into the modulation of inflammation by procyanidins will probably be tissue specific (with defined mechanisms of action being proposed), supporting beneficial effects gained from a diet high in fruit procyanidins. As research progresses, attention should be given to understanding the bioavailability and metabolism of individual procyanidin compounds and trying to identify them specifically in foods. Furthermore, understanding the mechanisms by which each procyanidin influences cellular signalling in more detail would allow for the focused use of single and multiple procyanidins in fresh and processed foods as a natural means to prevent inflammatory illness, limit the use of pharmaceutical intervention, and assist with improving human health.

References

[1] Xie, D. Y. and Dixon, R. A. Proanthocyanidin biosynthesis - still more questions than answers? *Phytochemistry* 66, 2127-2144 (2005).

[2] Traka, M. H. and Mithen, R. F. Plant Science and Human Nutrition: Challenges in Assessing Health-Promoting Properties of Phytochemicals. *Plant Cell* 23, 2483-2497 (2011).

[3] Zamora-Ros, R., et al. Dietary Intakes of Individual Flavanols and Flavonols Are Inversely Associated with Incident Type 2 Diabetes in European Populations. *Journal of Nutrition* 144, 335-343 (2014).

[4] Wang, J., et al. Targeting multiple pathogenic mechanisms with polyphenols for the treatment of Alzheimer's disease-experimental approach and therapeutic implications. *Frontiers in Aging Neuroscience* 6 (2014).

[5] Spector, W. G. *An Introduction to General Pathology*, Edn. Second. (Churchill Livingstone, New York; 1976).

[6] Holgate, S. T. The sentinel role of the airway epithelium in asthma pathogenesis. *Immunol. Rev.* 242, 205-219 (2011).

[7] Ishmael, F. T. The inflammatory response in the pathogenesis of asthma. *The Journal of the American Osteopathic Association* 111, S11-17 (2011).

[8] Murphy, D. M. and O'Byrne, P. M. Recent Advances in the Pathophysiology of Asthma. *Chest* 137, 1417-1426 (2010).

[9] Wu, Y., Antony, S., Meitzler, J. L., and Doroshow, J. H. Molecular mechanisms underlying chronic inflammation-associated cancers. *Cancer Letters* 345, 164-173 (2014).

[10] Feliciano, R. P., Meudt, J. J., Shanmuganayagam, D., Krueger, C. G., and Reed, J. D. Ratio of "A-type" to "B-type" Proanthocyanidin Interflavan Bonds Affects Extra-intestinal Pathogenic Escherichia coli Invasion of Gut Epithelial Cells. *J. Agric. Food Chem.* 62, 3919-3925 (2014).

[11] Mackenzie, G. G., Delfino, J. M., Keen, C. L., Fraga, C. G., and Oteiza, P. I. Dimeric procyanidins are inhibitors of NF-kappa B-DNA binding. *Biochem. Pharmacol.* 78, 1252-1262 (2009).

[12] Lu, Z., et al. Hypoglycemic activities of A- and B-type procyanidin oligomer-rich extracts from different Cinnamon barks. *Phytomedicine* 18, 298-302 (2011).

[13] Joven, J., et al. Polyphenols and the Modulation of Gene Expression Pathways: Can We Eat Our Way Out of the Danger of Chronic Disease? *Critical Reviews in Food Science and Nutrition* 54, 985-1001 (2014).

[14] Ou, K. and Gu, L. Absorption and metabolism of proanthocyanidins. *Journal of Functional Foods* 7, 43-53 (2014).

[15] Wang, Y., Chung, S.-J., Song, W. O., and Chun, O. K. Estimation of Daily Proanthocyanidin Intake and Major Food Sources in the US Diet. *Journal of Nutrition* 141, 447-452 (2011).

[16] Ovaskainen, M.-L., et al. Dietary intake and major food sources of polyphenols in Finnish adults. *Journal of Nutrition* 138, 562-566 (2008).

[17] Zamora-Ros, R., et al. Estimation of Dietary Sources and Flavonoid Intake in a Spanish Adult Population (EPIC-Spain). *Journal of the American Dietetic Association* 110, 390-398 (2010).

[18] Tulio, A. Z., Jr. et al. Phenolic composition, antioxidant properties, and endothelial cell function of red and white cranberry fruits. *Food Chemistry* 157, 540-552 (2014).

[19] Chen, Q., Yu, H., Tang, H., and Wang, X. Identification and expression analysis of genes involved in anthocyanin and proanthocyanidin biosynthesis in the fruit of blackberry. *Scientia Horticulturae* 141, 61-68 (2012).

[20] Gu, L. W., et al. Concentrations of proanthocyanidins in common foods and estimations of normal consumption. *Journal of Nutrition* 134, 613-617 (2004).

[21] Laboratory, N. D. (ed. A. R. S. Beltsville Human Nutrition Research Center, US Department of Agriculture) (US Department of Agriculture, 2004).

[22] Kviklys, D., et al. Rootstock genotype determines phenol content in apple fruits. *Plant Soil and Environment* 60, 234-240 (2014).

[23] Ceymann, M., Arrigoni, E., Schaerer, H., Nising, A. B., and Hurrell, R. F. Identification of apples rich in health-promoting flavan-3-ols and phenolic acids by measuring the polyphenol profile. *Journal of Food Composition and Analysis* 26, 128-135 (2012).

[24] Davies, K. M. and Espley, R. V. Opportunities and challenges for metabolic engineering of secondary metabolite pathways for improved human health characters in fruit and vegetable crops. *New Zealand Journal of Crop and Horticultural Science* 41, 154-177 (2013).

[25] Ricardo-Da-Silva, J. M., Rigaud, J., Cheynier, V., Cheminat, A., and Moutounet, M. Procyanidin dimers and trimers from grape seeds. *Phytochemistry* 30, 1259-1264 (1991).

[26] Santos-Buelga, C. and Scalbert, A. Proanthocyanidins and tannin-like compounds - nature, occurrence, dietary intake and effects on nutrition and health. *Journal of the Science of Food and Agriculture* 80, 1094-1117 (2000).

[27] Vogiatzoglou, A., et al. Assessment of the dietary intake of total flavan-3-ols, monomeric flavan-3-ols, proanthocyanidins and theaflavins in the European Union. *Br. J. Nutr.* 111, 1463-1473 (2014).

[28] Rzeppa, S., Von Bargen, C., Bittner, K., and Humpf, H.-U. Analysis of Flavan-3-ols and Procyanidins in Food Samples by Reversed Phase High-Performance Liquid Chromatography Coupled to Electrospray Ionization Tandem Mass Spectrometry (RP-HPLC-ESI-MS/MS). *J. Agric. Food Chem.* 59, 10594-10603 (2011).

[29] Gavrilova, V., Kajdzanoska, M., Gjamovski, V., and Stefova, M. Separation, Characterization and Quantification of Phenolic Compounds in Blueberries and Red and Black Currants by HPLC-DAD-ESI-MSn. *J. Agric. Food Chem.* 59, 4009-4018 (2011).

[30] Furuuchi, R., Yokoyama, T., Watanabe, Y., and Hirayama, M. Identification and Quantification of Short Oligomeric Proanthocyanidins and Other Polyphenols in Boysenberry Seeds and Juice. *J. Agric. Food Chem.* 59, 3738-3746 (2011).

[31] Jungfer, E., Zimmermann, B. F., Ruttkat, A., and Galensa, R. Comparing Procyanidins in Selected Vaccinium Species by UHPLC-MS2 with Regard to Authenticity and Health Effects. *J. Agric. Food Chem.* 60, 9688-9696 (2012).

[32] Tomas-Barberan, F. A., et al. HPLC-DAD-ESIMS analysis of phenolic compounds in nectarines, peaches, and plums. *J. Agric. Food Chem.* 49, 4748-4760 (2001).

[33] Nunes, C., et al. Characterization of plum procyanidins by thiolytic depolymerization. *J. Agric. Food Chem.* 56, 5188-5196 (2008).

[34] Wojdylo, A., Teleszko, M. and Oszmianski, J. Antioxidant property and storage stability of quince juice phenolic compounds. *Food Chemistry* 152, 261-270 (2014).

[35] Maatta-Riihinen, K. R., Kamal-Eldin, A. and Torronen, A. R. Identification and quantification of phenolic compounds in berries of Fragaria and Rubus species (Family rosaceae). *J. Agric. Food Chem.* 52, 6178-6187 (2004).

[36] Demir, N., Yildiz, O., Alpaslan, M., and Hayaloglu, A. A. Evaluation of volatiles, phenolic compounds and antioxidant activities of rose hip (Rosa L.) fruits in Turkey. *LWT-Food Sci. Technol.* 57, 126-133 (2014).

[37] Velderrain-Rodriguez, G. R., et al. Phenolic compounds: their journey after intake. *Food Funct.* 5, 189-197 (2014).

[38] Pierre, J. F., et al. Cranberry Proanthocyanidins Improve Intestinal sIgA During Elemental Enteral Nutrition. *J. Parenter. Enter. Nutr.* 38, 107-114 (2014).

[39] Serra, A., et al. Determination of procyanidins and their metabolites in plasma samples by improved liquid chromatography-tandem mass

spectrometry. *Journal of Chromatography B-Analytical Technologies in the Biomedical and Life Sciences* 877, 1169-1176 (2009).

[40] Appeldoorn, M. M., Vincken, J.-P., Gruppen, H., and Hollman, P. C. H. Procyanidin Dimers A1, A2, and B2 Are Absorbed without Conjugation or Methylation from the Small Intestine of Rats. *Journal of Nutrition* 139, 1469-1473 (2009).

[41] Schantz, M., Erk, T. and Richling, E. Metabolism of green tea catechins by the human small intestine. *Biotechnology Journal* 5, 1050-1059 (2010).

[42] Tanaka, N., et al. Measurement of plasma procyanidin B-2 and procyanidin B-3 levels after oral administration in rat. *Phytomedicine* 10, 122-126 (2003).

[43] Serra, A., Blade, C., Arola, L., Macia, A., and Motilva, M. J. Flavanol metabolites distribute in visceral adipose depots after a long-term intake of grape seed proanthocyanidin extract in rats. *Br. J. Nutr.* 110, 1411-1420 (2013).

[44] Manach, C., Williamson, G., Morand, C., Scalbert, A., and Remesy, C. Bioavailability and bioefficacy of polyphenols in humans. I. Review of 97 bioavailability studies. *The American journal of clinical nutrition* 81, 230S-242S (2005).

[45] Verstraeten, S. V., Jaggers, G. K., Fraga, C. G., and Oteiza, P. I. Procyanidins can interact with Caco-2 cell membrane lipid rafts: Involvement of cholesterol. *Biochimica Et Biophysica Acta-Biomembranes* 1828, 2646-2653 (2013).

[46] Grivennikov, S. I. and Karin, M. Dangerous liaisons: STAT3 and NF-kappa B collaboration and crosstalk in cancer. *Cytokine and Growth Factor Reviews* 21, 11-19 (2010).

[47] Blade, C., Arola, L. and Salvado, M.-J. Hypolipidemic effects of proanthocyanidins and their underlying biochemical and molecular mechanisms. *Mol. Nutr. Food Res.* 54, 37-59 (2010).

[48] Kawaguchi, K., Matsumoto, T. and Kumazawa, Y. Effects of Antioxidant Polyphenols on TNF-Alpha-Related Diseases. *Current Topics in Medicinal Chemistry* 11, 1767-1779 (2011).

[49] Krajka-Kuzniak, V., Paluszczak, J., Oszmianski, J., and Baer-Dubowska, W. Hawthorn (Crataegus oxyacantha L.) Bark Extract Regulates Antioxidant Response Element (ARE)-Mediated Enzyme Expression Via Nrf2 Pathway Activation in Normal Hepatocyte Cell Line. *Phytotherapy Research* 28, 593-602 (2014).

[50] Nguyen, T., Nioi, P. and Pickett, C. B. The Nrf2-Antioxidant Response Element Signaling Pathway and Its Activation by Oxidative Stress. *Journal of Biological Chemistry* 284, 13291-13295 (2009).

[51] Reuter, S., Gupta, S. C., Chaturvedi, M. M., and Aggarwal, B. B. Oxidative stress, inflammation, and cancer How are they linked? *Free Radical Biology and Medicine* 49, 1603-1616 (2010).

[52] Kumar, A., Takada, Y., Boriek, A. M., and Aggarwal, B. B. Nuclear factor-kappa B: its role in health and disease. *Journal of Molecular Medicine-Jmm* 82, 434-448 (2004).

[53] Sarkar, F. H., Li, Y., Wang, Z., and Kong, D. NF-kappa B Signaling Pathway and Its Therapeutic Implications in Human Diseases. *International Reviews of Immunology* 27, 293-319 (2008).

[54] Martinez-Micaelo, N., Gonzalez-Abuin, N., Ardevol, A., Pinent, M., and Blay, M. T. Procyanidins and inflammation: Molecular targets and health implications. *Biofactors* 38, 257-265 (2012).

[55] Zenz, R., et al. Activator protein 1 (Fos/Jun) functions in inflammatory bone and skin disease. *Arthritis Research and Therapy* 10 (2008).

[56] Tyagi, A., et al. Procyanidin B2 3,3"-di-O-gallate, a Biologically Active Constituent of Grape Seed Extract, Induces Apoptosis in Human Prostate Cancer Cells Via Targeting NF-kB, Stat3, and AP1 Transcription Factors. *Nutrition and Cancer-an International Journal* 66, 736-746 (2014).

[57] Takano, F., et al. Aqueous extract of peanut skin and its main constituent procyanidin Al suppress serum IgE and IgG1 levels in mice-immunized with ovalbumin. *Biological and Pharmaceutical Bulletin* 30, 922-927 (2007).

[58] Karamouzis, M. V., Konstantinopoulos, P. A. and Papavassiliou, A. G. The activator protein-1 transcription factor in respiratory epithelium carcinogenesis. *Mol. Cancer Res.* 5, 109-120 (2007).

[59] Martinez-Micaelo, N., et al. Omega-3 docosahexaenoic acid and procyanidins inhibit cyclo-oxygenase activity and attenuate NF-kappa B activation through a p105/p50 regulatory mechanism in macrophage inflammation. *Biochemical Journal* 441, 653-663 (2012).

[60] Hurst, S. M., et al. Blackcurrant proanthocyanidins augment IFN-gamma-induced suppression of IL-4 stimulated CCL26 secretion in alveolar epithelial cells. *Mol. Nutr. Food Res.* 54, S159-S170 (2010).

[61] Aittomaki, S. and Pesu, M. Therapeutic Targeting of the JAK/STAT Pathway. *Basic and Clinical Pharmacology and Toxicology* 114, 18-23 (2014).

[62] Vahedi, G., et al. STATs Shape the Active Enhancer Landscape of T Cell Populations. *Cell* 151, 981-993 (2012).

[63] Yang, D. J., et al. Inhibitory Effect of Litchi (Litchi chinensis Sonn.) Flower on Lipopolysaccharide-Induced Expression of Proinflammatory Mediators in RAW264.7 Cells through NF-kappa B, ERK, and JAK2/STAT3 Inactivation. *J. Agric. Food Chem.* 62, 3458-3465 (2014).

[64] Andlujar, I., et al. Inhibition of Ulcerative Colitis in Mice after Oral Administration of a Polyphenol-Enriched Cocoa Extract Is Mediated by the Inhibition of STAT1 and STAT3 Phosphorylation in Colon Cells. *J. Agric. Food Chem.* 59, 6474-6483 (2011).

[65] Berger, A. *Th1 and Th2 responses: what are they?*, Vol. 321. (2000).

[66] Connor, C. A., Adriaens, M., Pierini, R., Johnson, I. T., and Belshaw, N. J. Procyanidin Induces Apoptosis of Esophageal Adenocarcinoma Cells via JNK Activation of c-Jun. *Nutrition and Cancer-an International Journal* 66, 335-341 (2014).

[67] Terra, X., et al. Procyanidin dimer B1 and trimer C1 impair inflammatory response signalling in human monocytes. *Free Radical Research* 45, 611-619 (2011).

[68] Passos, P. C. (ed.) *Grape Seed Procyanidins: Occurrence, Structural Features, and Health Benefits.* (Nova Science Publishers, Incorporated, 2010).

[69] Orton, R. J., et al. Computational modelling of the receptor-tyrosine-kinase-activated MAPK pathway. *Biochemical Journal* 392, 249-261 (2005).

[70] Roskoski, R., Jr. ERK1/2 MAP kinases: Structure, function, and regulation. *Pharmacological Research* 66, 105-143 (2012).

In: Proanthocyanidins
Editor: Ingrid Sullivan

ISBN: 978-1-63463-430-4

Chapter 4

DIETARY ANTHOCYANINS: POTENTIAL NEUROPROTECTIVE AGENTS IN NEURODEGENERATIVE DISEASES?

Eduardo Luiz Gasnhar Moreira[1,*], Manuel Adalberto Álfaro De Prá[1], Fausto Rodrigues Cardoso[1], Andreza Fabro de Bem[2], Rosa Maria Ribeiro-do-Valle Nicolau[1] and Marcelo Maraschin[1]

[1] Plant Morphogenesis and Biochemistry Laboratory, Core of Natural Produtcs, Federal University of Santa Catarina, Florianópolis, SC, Brasil

[2] Department of Biochemistry, Biological Science Center, Federal University of Santa Catarina, Florianópolis, SC, Brasil

ABSTRACT

Neurodegenerative disorders such as Parkinson's disease (PD) and Alzheimer's disease (AD) represent increasing problems in our aging societies, mostly with an increased prevalence of both diseases with age. Evidence suggests that lifestyle factors, especially nutrition, can influence the incidence and onset of neurodegenerative diseases and thus a healthy diet is an essential factor for health aging. Anthocyanins are the largest

* Correspondence: Eduardo Luiz Gasnhar Moreira, PhD; E-mail : eduardoluiz26@gmail.com.

group of water-soluble pigments in the plant kingdom and belong to the family of secondary metabolites known as flavonoids. Major sources of anthocyanins are blueberries, cherries, raspberries, strawberries, black currants, purple grapes, and red wine. Anthocyanin pigments and associated flavonoids have demonstrated ability to protect against a myriad of human diseases. In particular, several studies have shown that the consumption of diets rich in antioxidant components, such as anthocyanins, may modulate the progression of neurodegenerative diseases such as AD and PD. This chapter is aimed at presenting the neuroprotective effects of anthocyanins and their interactions with specific molecular targets and pathways of neurodegeneration process. Potentials and pitfalls of dietary anthocyanins as neuroprotective agents have been discussed as well.

Keywords: Anthocyanins; Parkinson's disease; Alzheimer's disease; Neurodegenerative diseases; Natural antioxidants

1. Introduction

Neurodegenerative diseases result from the gradual and progressive loss of neural cells, leading to nervous system dysfunction [1]. These diseases are characterized by an abnormal accumulation of oxidatively damaged macromolecules inside cells and in the extracellular space [2]. Proteins involved in the formation of aggregates are α-synuclein, parkin, β-amyloid, prion proteins, and proteins containing polyglutamine [2]. Protein aggregates subsequently lead to the toxic disruption of normal cellular metabolism leading, ultimately, to disease [3]. Neurological signs and symptoms are markedly varied, but the commonest neurodegenerative diseases can be classified according to their consequences, e.g., cause dementia, movement disorders, or motor disorders. The most common neurodegenerative diseases are Alzheimer´s disease (AD), which is the major cause of dementia, and Parkinson´s disease (PD), which is a movement disorder. To date, as no curative treatments exist for these disorders it is incumbent upon future research efforts to focus on understanding the key elements behind these disorders [4].

As with other slowly progressive diseases, preventing neurodegenerative diseases depends on understanding early steps in its pathogenesis [5]. The etiology of neurodegenerative disorders is multifactorial and consists of an interaction among aging, environmental factors, and genetic predisposition [5-6]. One of the main characteristics of neurodegenerative disease is the selective vulnerability of different neuronal populations that are affected in a progressive and commonly stereotyped manner [7]. Specifically, the most vulnerable regions of the brain affected in AD such as the entorhinal cortex and the hippocampus CA1 region differ from the most vulnerable regions affected in PD which includes the dopaminergic neurons in the substantia nigra pars compacta [7]. However, although both disorders have distinct pathologic features, a great amount of evidence points to oxidative stress as a central mechanism of chronic neurodegeneration [8]. High oxygen consumption, high production of reactive oxygen species (ROS), abundant oxidation-sensitive lipids, low antioxidant defense, and a restricted renewal and regenerative capacity of neurons render the brain especially susceptible to oxidative insults [9]. Oxidative stress is intimately linked with an integrated series of cellular phenomena, which all seem to contribute to neuronal demise [10]. In particular, Andersen (2004) proposed that the interaction between these various components is not necessarily a cascade, but might be a cycle of events of which oxidative stress is a major component [10]. It is worth noting that oxidation protein modification occurs at a low and persistent level in diverse cells and tissues and accumulates in aging and neurodegenerative diseases [11].

Based on the hypothesis that oxidative stress is a crucial player in common neurodegenerative diseases, the rationale for the use of antioxidants as disease-modifying agents is clear. Not surprisingly, studies of *in vitro* models of neuronal toxicity and animal models of neurodegenerative disorders have demonstrated potential neuroprotective effects of antioxidant therapies. However, despite the promise of these animal studies, human clinical experience with antioxidant neuroprotectants has been generally negative [12]. With respect to nutritional supplement studies most human data have been collected from vitamin E trials involving mixtures of vitamin antioxidants and micronutrients [12]. Accordingly to Fusco et al. (2007), a major limitation present in most of the intervention studies exploring the effects of antioxidants supplementation (e.g., vitamin E) on AD and PD outcomes is that they have been conducted on subjects who already have been diagnosed with the clinical condition [13]. Thus, it is difficult to assess the full potential of the specific substances on the prevention of AD and PD. Moreover, it is hypothesized that a combination of antioxidants is more effective than a single antioxidant, due to synergistic effects of compounds.

On the other hand, modification of nutritional habits among other lifestyle changes might reduce the risk for developing neurodegenerative diseases in later life. For instance, epidemiological studies indicated that dietary habits can influence the incidence of neurodegenerative disorders such as AD [14-15] and PD [16-17]. In this context, a higher daily intake of fruits and vegetables is associated with an improved antioxidant status in comparison to subjects consuming diets poor in fruits and vegetables [18]. These effects are most likely due to the presence of plant polyphenols [19], a large group of natural antioxidants ubiquitous in a diet rich in vegetables and fruits. Of note, epidemiological studies have shown an inverse association between the consumption of polyphenol-rich foods or beverages and the risk of chronic diseases, such as cancer, cardiovascular diseases, and neurodegenerative diseases [16, 20-23]. Of particular interest herein, in the last decade literature on the neuroprotective effects of a polyphenol-rich diet has grown considerably. Specifically, a large variety of plant polyphenols exist, including benzoic acids, cinnamic acids, stilbenes, coumarins, lignins, lignans, and flavonoids [24].

Anthocyanins are water soluble flavonoid pigments, widely spread throughout the plant kingdom, that are particularly abundant in berries and other fruits with red, blue, or purple color, and in red wines [25]. The daily intake of anthocyanins in humans has been estimated at 180-215 mg/day [26]. However, anthocyanins can be major dietary components with intakes in excess of 1 g being feasible for those who routinely eat berries and drink red wine [27]. Interest in the biological effects of anthocyanin compounds has increased substantially during the last decade because of increasing evidence demonstrating potential therapeutic effects. In particular, some studies have shown that the consumption of diets rich in antioxidant components, such as anthocyanins, may modulate the progression of neurodegenerative diseases such as AD and PD [17, 28].

In this chapter, we have focused on neuroprotective effects of anthocyanins and their interaction with specific molecular targets and pathways of neurodegeneration. Furthermore, the potentials and pitfalls of dietary anthocyanins as neuroprotective agents have also been discussed.

2. Anthocyanins: Sources and Structure

Anthocyanins are water-soluble glycosides of polyhydroxy and polymethoxy derivatives of 2-phenylbenzopyrylium or flavylium salts [25].

They belong to the widespread class of phenolic compounds collectively named flavonoids, possessing a characteristic C_3–C_6–C_3 carbon structure [25]. Anthocyanins occur naturally in plants in the form of glycosides in which the anthocyanidin molecule is coupled to a sugar [29]. Glycosylation confers increased structural stability and water solubility to the parent anthocyanidin (aglycone). Anthocyanins are responsible for the orange-to-blue colors found in many flowers, leaves, fruits, seeds, and other tissues [30]. In this context, they are believed to be important to plants as their color attracts animals, leading to seed dispersal and pollination [31]. Moreover, anthocyanins are a protective mechanism of plants against environmental stress factors such as UV light, cold temperatures, and drought [32].

To date, there have been more than 635 anthocyanins identified in nature [31-32]. The six anthocyanidins commonly found in fruit and vegetables are cyanidin, delphinidin, malvidin, peonidin, pelargonidin, and petunidin which are combined with monosaccharides, mostly glucose, arabinose, and galactose [33]. The differences between individual anthocyanins are related to the number and the position of hydroxyl groups, the degree of methylation of these hydroxyl groups; the aliphatic or aromatic acids attached to the sugars in the molecule; and the nature, number, and location of sugars attached to the molecule [25]. Moreover, the chromophore of 8 conjugated double bonds carrying a positive charge on the heterocyclic oxygen ring is responsible for the intense red-orange to blue-violet color produced by anthocyanins under acidic conditions [32].

C. Brain Bioavailability of Anthocyanins

Previous studies on antioxidants as neuroprotective agents have been hampered by the impermeability of the blood-brain barrier (BBB) to many compounds. Thus, any neuroprotective effects that anthocyanins might exert are dependent on sufficient brain bioavailability through the diet. In this context, Milbury and Kalt (2010) demonstrated that Jersey blueberry anthocyanins when fed to pigs are bioavailable, transit the BBB, and can be found in brain tissue [34]. Specifically, anthocyanins and their glucoronides were found in the range of femtomoles per gram of fresh weight of tissue at 18 h postprandial, after anthocyanins had been removed from the blood by xenobiotic metabolism [34]. Moreover, Andres-Lacueva et al. (2005) demonstrated that following 10-weeks of dietary supplementation with 2% blueberries, anthocyanins were found in the brains of all rats fed blueberries [35]. In particular, the galactoside forms of cyaniding and malvidin and malvidin-3-*O*-β-arabinose were identified in a number of brain regions, most frequently in the hippocampus and the cortex [35]. In accordance, Rashid et al. (2013) also provided evidence that Kenyan purple tea anthocyanins are able to cross the BBB and exert their physiological effects in this organ [36]. Talavera et al. (2005) demonstrated that anthocyanins target the brain of rats following consumption of an anthocyanin-enriched diet for 15 days (14.8 mmol anthocyanins from blackberry per kg diet) [37]. Specifically, the authors demonstrated that cyaniding 3-glucoside content was higher in the brain (0.20 ± 0.05 nmol/g of tissue) than in plasma (0.15 ± 0.02 nmol/mL). Moreover, the authors reported that a single oral dose of blackberry anthocyanins administered by gavage has shown that anthocyanins were present in the brain only 30 min after their administration [37]. Interestingly, in pig cerebellum, anthocyanin concentration did not rise uniformly, which implies that anthocyanin absorption may become saturated at high dosage levels [38].

However, although several studies have indicated that absorbed anthocyanins and their metabolites are able to cross the BBB, direct antioxidant effects of anthocyanins are not very likely to occur in the brain *in vivo*, especially as anthocyanins *in vivo* are likely found in too low contents to matter in the face of endogenous antioxidants [39]. On the other hand, anthocyanins and other polyphenols may exert neuropharmacological actions at the molecular level, influencing signaling pathways, gene expression, and protein function [40-42], as discussed below.

D. Biological Activities of Anthocyanins

Numerous studies have shown that anthocyanins display a wide range of biological activities including antioxidant [43-44], anti-inflammatory [45], antimicrobial [46], and anti-carcinogenic [47-48]. In addition, anthocyanins display a variety of effects on blood vessels and platelets that may reduce the risk of coronary heart disease [49-51]. Anthocyanins are potent antioxidant superior to classical antioxidants such as butylated hydroxyanisole (BHA), butylated hydroxytoulene (BHT), and α-tocopherol [43, 52]. For instance, Mazza et al. (2002) found that the concentration of anthocyanins in the serum of male subjects that consumed 1.2 g anthocyanins from freeze dried blueberries was positively correlated with the serum antioxidant capacity [44]. However, although the biological effects of anthocyanins have historically been attributed to their antioxidant actions, the classic hydrogen-donating antioxidant activity may not account for their *in vivo* bioactivity, particularly in the brain where they are found in low concentrations [39]. In this regard, other relevant potential mechanisms include interactions with neuronal signaling pathways that are critical in controlling neuronal survival and differentiation and in modulating activity/expression of several oxidative-related enzymes, and regulation of mitochondrial function or neuroinflammation as discussed below.

For instance, Lee et al. (2013) showed that the anti-inflammatory effect of the berry anthocyanin fractions were, at least partly, attributed to the inhibition of the nuclear translocation of the nuclear factor κB (Nf-κB), a master transcription factor that induces the expression of proinflammatory cytokines [42]. The same authors found that the activation of the transcription of the NF-E2-related factor 2 (Nrf2) plays a critical role in eliciting the antioxidant effects of berry anthocyanin fractions by reducing cellular ROS levels [42]. In this context, Lu et al. (2010) showed that dietary supplementation with purple sweet potato color, a class of naturally occurring anthocyanins, improved the learning and memory ability of D-galactose-treated old mice, accompanied with increased expression and activity of Cu, Zn-superoxide dismutase and catalase [53].

In a very elegant study, Wiliams et al. (2008) showed that a supplementation with a blueberry diet (2% w/w) for 12 weeks improves the performance of aged animals in spatial working memory tasks [40]. Memory performance correlated well with the activation of cAMP-response element-binding protein (CREB) and increases in both pro- and mature levels of brain-derived neurotrophic factor (BDNF) in the hippocampus. Changes in CREB and BDNF in aged and blueberry-supplemented animals were accompanied by increases in the phosphorylation state of extracellular signal-related kinase (ERK1/2) [40]. Besides, age and blueberry supplementation were linked to changes in the activation state of Akt (protein kinase B) and mTOR (mechanistic target of rapamycin) in the hippocampus, suggesting that pathways involved in *de novo* protein synthesis may be involved [40].

Moreover, Dreiseitel et al. (2009) examined the impact of anthocyanins on monoamine oxidases (MAO A and MAO B) activities [54]. Kinetic analyses performed with cyaniding and cyaniding-3-glucoside indicated a competitive interaction of cyaniding with MAO A plus a mixed competitive and non-competitive mode of interaction of cyaniding with MAO B and of cyaniding-3-glucoside with both enzyme isoforms. According to the authors, anthocyanins and their aglycons achieved MAO inhibition *in vitro* that is compatible with central nervous functionalities [54]. It is worth mentioning that in clinical practice selective inhibitors of MAO B are routinely used to alleviate symptoms or to slow the progression of PD [55]. Furthermore, Maeda-Yamamoto et al. (2012) showed that an anthocyanin-rich tea cultivar 'Sunrouge' (*Camellia sinensis x C. taliensis*) effectively inhibited the acetylcholinesterase (AChE) activity in human neuroblastoma SK-N-SH cells [56]. Importantly, cholinesterase inhibitors are widely used in the management of cognitive impairments and dementia [57]. Overall, the above mentioned properties of anthocyanins allow them to afford neuroprotection in different neurodegenerative disorders as further discussed.

E. Basic Studies

There is growing experimental evidence suggesting neuroprotective effects of anthocyanin-rich extracts, although they do not represent necessarily dietary relevant components. Moreover, although such studies have significantly contributed to the elucidation of some of the underlying mechanisms mediating the neuroprotection elicited by anthocyanins, they are scarce so far.

Qin et al. (2013) evaluated the effects of cyaniding 3-*O*-glucoside in an Aβ-induced model of AD [58]. Male Sprague-Dawley rats received 5μl of Aβ 1-42 (10μg/μl) injected bilaterally into the hippocampus using a 26 gauge needle fixed to a Hamilton syringe. After that, cyaniding 3-*O*-glucoside was administered orally via syringe once daily for 30 days at a dose of 10mg/kg. The authors found that cyaniding 3-*O*-glucoside is protective against the cognitive impairments that are induced by the infusion of Aβ peptide via the modulation of GSK-3β/tau signaling pathway [58]. In this context, Shih et al. (2010) evaluated the beneficial effect of mulberry extracts rich in anthocyanins on the induction of antioxidant enzymes and on the promotion of cognition in senescence-accelerated mice (SAMP) [59]. In this study, six-month old SAMP8 and SAMR1 mice were fed a basal diet supplemented with 0.18% and 0.9% (w/w) mulberry extracts for 12 consecutive weeks. The results showed that the mice fed with mulberry extracts supplement presented significantly less amyloid β protein and showed improved learning and memory ability in avoidance response tests. The protective effect of mulberry extracts appeared to occur through a mechanism related to regulation of the mitogen-activated protein kinases and activation of the Nrf2 [59], where the latter regulates the induction of phase 2 antioxidant enzymes and reduction of oxidative damage. In their turn, Vepsäläinen et al. (2013) investigated the effects of anthocyanin-rich blackcurrant supplementation diet (i.e., total amount of anthocyanins = 1.43 mg/g) on Aβ and tau pathology, as well as on the behavioral deficits in the APdE9 mouse model of AD (mice overexpressing mutated human APP and presenilin-1 proteins) [60]. The authors started the dietary intervention at the age of 2.5 months before any Aβ plaque deposition is detectable in the brain of the mice, continuing the study until the age of 13 months when mice showed clear memory deficits and massive Aβ deposition in the brain. APdE9 mice fed with anthocyanin-rich blackcurrant extract showed decreased APP C-terminal fragment levels in the cerebral cortex as compared to APdE9 mice on the control diet. The diet also alleviated the spatial working memory deficit of aged APdE9 mice as compared to mice fed control diet [60].

Kim et al. (2010) examined the effects of the ethanol extract of mulberry fruit in an *in vivo* model of PD induced by 1-methyl-4-phenyl-1,2,3,6-tetrahydropyridine (MPTP) [61]. Mulberry extract was administered at 500 mg/kg per day for 15 days. MPTP was injected at 30 mg/kg (i.p.) per day for the last 5 days of Mulberry treatment. The authors found that the ethanol extract of mulberry fruit showed preventative effect against PD-like symptoms (bradykinesia) in the behavioral test and prevented MPTP-induced dopaminergic neuronal damage in an immuno-cytochemical analysis of the substantia nigra pars compacta and striatum [61]. Of note, our research group demonstrated that a neuroactive proanthocyanidin-rich fraction obtained from the bark of *Croton celtidifolius* Baill [62] confers neuroprotection in the intranasal MPTP rat model of PD [63]. Specifically, it was shown that the pre-treatment with the proanthocyanidin-rich fraction (10mg/kg, i.p.) during five consecutive days was able to prevent mitochondrial complex-I inhibition in the striatum and olfactory bulb, as well as a decrease of the enzyme tyrosine hydroxylase expression in the olfactory bulb and substantia nigra of rats infused with a single intranasal administration of MPTP (1mg/nostril – [63]). Additionally, we recently also demonstrated that this proanthocyanidin-rich fraction confers neuroprotection in an experimental model of spinal cord injury, event mediated at least partly by the inhibition of N-methyl-D-aspartate-type glutamate receptors [64].

In a very elegant study, Strathearn et al. (2014) characterized a number of botanical extracts rich in anthocyanins and proanthocyanidins in terms of their neuroprotective activities in primary midbrain cultures exposed to rotenone. Specifically, dopaminergic cell death elicited by rotenone was suppressed by extracts prepared from blueberries, grape seed, hibiscus, blackcurrant, and Chinese mulberry. Extracts rich in anthocyanins and proanthocyanidins exhibited greater neuroprotective activity than extracts rich in other polyphenols, and a number of individual anthocyanins such as the malvidin-3-O-glucoside interfered with rotenone neurotoxicity [65].

F. Human Studies

The majority of clinical research regarding dietary anthocyanins has focused on their cardioprotective effects. In fact, epidemiological studies suggest that the daily intake of berry anthocyanins may have cardioprotective effects in humans [66-68]. On the other hand, although evidence supporting the neuroprotective effects of anthocyanins have increased significantly in recent years, to date much of the findings have emerged from animal rather human studies. In this regard, Gao et al. (2012) carried out the first prospective study to examine whether greater intakes of total flavonoid or their sub-classes (i.e., flavanones, anthocyanins, flavan-3-ols, flavonols, flavones, and polymers) lowered the risk of developing PD [17]. In this large prospective study, the authors found that participants with greater consumption of anthocyanins, quercetin, epicatechin, and some proanthocyanidins showed to be lesser prone to develop PD during 20-22 years of follow-up. Higher intakes of anthocyanin-rich foods, such as berries, were associated with lower PD risk [17]. Previously, in a case-control study, Golbe et al. (1988) observed that PD patients (n = 81) were found to be 40% - 60% less likely to consume blueberries or strawberries compared to controls [69].

A few trials in older human subjects (n < 15 participants; participants older than 75 years old) have revealed improved short-term cognitive performance with anthocyanin-rich fruit juices. For instance, Krikorian et al. (2010), investigated the effects of daily consumption of wild blueberry juice in a sample of nine older adults with early memory changes [70]. At 12 weeks, improved paired associated learning and word list recall were observed. In addition, there were trends suggesting reduced depressive symptoms and lower glucose levels in blood plasma [70]. The same authors enrolled twelve older adults with memory decline, but not dementia, in a randomized, placebo-controlled, double-blind trial with Concord grape juice supplementation for 12 weeks [71]. They were able in detecting significant improvement in a measure of verbal learning [71]. In another study of the same research group, older adults subjects with mild cognitive impairment consumed Concord grape juice or placebo for 16 weeks and were administered assessments of memory function and brain activation pre- and post-intervation. Participants who consumed grape juice showed reduced semantic interference on memory tasks. Relatively greater activation in anterior and posterior regions of the right hemisphere was also observed with functional magnetic resonance imaging in the grape juice-treated subjects [72].

The study by Devore et al. (2012) was the first to evaluate the associations of long-term dietary intake of berries and anthocyanidins, with cognitive decline in a large, prospective cohort [28]. The study was conducted with older women of the Nurses' Health Study (participants older than 70 years old). Higher, long-term consumption of berries, anthocyanidins, and total dietary flavonoids were related to significantly slower rates of cognitive decline in this cohort of older women, even after careful consideration of confounding by socioeconomic status. The authors reported the first epidemiologic evidence that greater intakes of blueberries and strawberries (top food contributors to anthocyanidin intake) were highly associated with slower rates of cognitive decline [28].

In despite of the increasing body of evidences of the effects of anthocyanins on the cognitive response of human subjects, currently it seems consensus that more research is needed to determine the most efficacious intake levels in humans and to explore the potential for overconsumption of berry fruits. Besides, there are numerous studies indicating that a moderate consumption of red wine provides certain health benefits, such as the protection against neurodegenerative diseases [73-74]. For instance, Orgogozo et al. (1997) have shown a positive correlation between a moderate consumption of red wine and a decreased incidence of dementia [75]. Of particular interest herein, anthocyanins are the main phenolic compounds involved in the color of red wines [76]. However, it is worth mentioning that in addition to anthocyanins, red wine contains a broad range of polyphenols including phenolic acids, stilbenes, tannins, flavanols and flavonols [77]. In particular, the stilbene resveratrol has been one of the most extensively studied nonflavonoid as a critical constituent that contributes to the health benefits of red wine [78].

Conclusion

As life spans in many countries continue to increase, primary and secondary prevention of neurodegenerative diseases at the individual and population levels might diminish the future increase in the prevalence of these disorders. Prevention is particularly important in neurodegenerative diseases as there is no disease-modifying treatment available at present. In this regard, diets rich in fruits and vegetables have been shown to improve human well-being and to significantly delay the development of pathologic processes [79-82], including neurodegenerative disorders [14, 83]. Thus, a deep understanding on the beneficial health effects of diets rich in fruits and vegetables is imperative and, in the long run, may contribute to a healthier aging process and decreased prevalence of common neurodegenerative diseases.

In this context, the understanding of the chemistry, bioavailability, and biological activity of anthocyanins continues to evolve. Although the mechanisms by which anthocyanins act in the brain are not clear, there is evidence to suggest that anthocyanins can cross the BBB and reach the central nervous system, where they have the potential to directly regulate gene and protein expression in neurons. Anthocyanins are thought to work by modulating cellular signaling pathways, such as the MAPK and the phosphoinositide 3-kinase (PI3 kinase/Akt) signaling cascade [59]. They are also know to modulate the transcription factor NF-κB [42], which is involved in signal transduction through protein-kinase inhibition, and to promote the expression of BDNF [40-41], which is responsible for neuronal and synaptic growth and survival. The elevation of hippocampal protein levels of BDNF by anthocyanins is of particular importance because (*i*) BDNF expression in the human hippocampus is known to decrease with age [84] and (*ii*) BDNF is a key regulator of activity-dependent synaptic plasticity [85-87].

Since the pathological processes of diseases of neurodegenerative nature begin years, if not decades, before the clinical diagnosis [88-89], the efforts should be directed towards the prevention of symptoms early in life [90]. Thus, the daily intake of dietary anthocyanins holds great promise in modulating the development and progression of neurodegenerative diseases. Nevertheless, while a growing number of studies point to dietary anthocyanins as neuroprotective agents, more long-term, randomized, controlled, intervention trials with appropriate controls are required in order to assess the full and unequivocal role that anthocyanins play in preventing neurodegenerative diseases. The outcomes of these studies will provide a basis for more specific dietary, even though food supplements, and to fully validate anthocyanins as disease-modifying agents in neurodegenerative conditions.

Acknowledgments

This work was supported by grants from the Brazilian Institution Conselho Nacional de Desenvolvimento Científico e Tecnológico (CNPq) [402539/2013-3; MCTI/CNPq/CT-Biotec – Agregação de valores à agrobiodiversidade vegetal por rota biotecnológica]. ELGM, MAP, AFB, RMR, MM are supported by research fellowships from CNPq-Brazil. The authors have no financial or personal conflicts of interested related to this work.

References

[1] Brown, RC; Lockwood, AH; Sonawane BR. Neurodegenerative diseases: an overview of environmental risk factors. *Environ Health Perspect,* 2005 113 (9):1250-1256.

[2] Grimm, S; Hoehn, A; Davies, KJ; Grune T. Protein oxidative modifications in the ageing brain: consequence for the onset of neurodegenerative disease. *Free Radic Res,* 2011 45 (1):73-88.

[3] Weihl, CC. Monitoring autophagy in the treatment of protein aggregate diseases: steps toward identifying autophagic biomarkers. *Neurotherapeutics,* 2013 10 (3):383-390.

[4] Montie, HL; Durcan, TM. The cell and molecular biology of neurodegenerative diseases: an overview. *Front Neurol,* 2013 4:194.

[5] Selkoe, DJ. Preventing Alzheimer's disease. *Science,* 2012 337 (6101):1488-1492.

[6] Bertram, L; Tanzi, RE. The genetics of Alzheimer's disease. *Prog Mol Biol Transl Sci,* 2012 107:79-100.

[7] Gandhi, S; Abramov, AY. Mechanism of oxidative stress in neurodegeneration. *Oxid Med Cell Longev,* 2012 2012:428010.

[8] Lin, MT; Beal, MF. Mitochondrial dysfunction and oxidative stress in neurodegenerative diseases. *Nature,* 2006 443 (7113):787-795.

[9] Gao, HM; Zhou, H; Hong, JS. NADPH oxidases: novel therapeutic targets for neurodegenerative diseases. *Trends Pharmacol Sci*, 2012 33 (6):295-303.

[10] Andersen, JK. Oxidative stress in neurodegeneration: cause or consequence? *Nat Med,* 2004 10 Suppl:S18-25.

[11] Giordano, S; Darley-Usmar, V; Zhang, J. Autophagy as an essential cellular antioxidant pathway in neurodegenerative disease. *Redox Biol,* 2014 2:82-90.

[12] Kamat, CD; Gadal, S; Mhatre, M; Williamson, KS; Pye, QN; Hensley, K. Antioxidants in central nervous system diseases: preclinical promise and translational challenges. *J Alzheimers Dis,* 2008 15 (3):473-493.

[13] Fusco, D; Colloca, G; Lo Monaco, MR; Cesari, M. Effects of antioxidant supplementation on the aging process. *Clin Interv Aging,* 2007 2 (3):377-387.

[14] Scarmeas, N; Stern, Y; Mayeux, R; Luchsinger, JA. Mediterranean diet, Alzheimer disease, and vascular mediation. *Arch Neurol,* 2006 63 (12):1709-1717.

[15] Sofi, F; Macchi, C; Abbate, R; Gensini, GF; Casini, A. Effectiveness of the Mediterranean diet: can it help delay or prevent Alzheimer's disease? *J Alzheimers Dis,* 2010 20 (3):795-801.

[16] Costa, J; Lunet, N; Santos, C; Santos, J; Vaz-Carneiro, A. Caffeine exposure and the risk of Parkinson's disease: a systematic review and meta-analysis of observational studies. *J Alzheimers Dis*, 2010 20 Suppl 1:S221-238.

[17] Gao, X; Cassidy, A; Schwarzschild, MA; Rimm, EB; Ascherio, A. Habitual intake of dietary flavonoids and risk of Parkinson disease. *Neurology,* 2012 78 (15):1138-1145.

[18] Anlasik, T; Sies, H; Griffiths, HR; Mecocci, P; Stahl, W; Polidori, MC. Dietary habits are major determinants of the plasma antioxidant status in healthy elderly subjects. *Br J Nutr,* 2005 94 (5):639-642.

[19] Esposito, E; Rotilio, D; Di Matteo, V; Di Giulio, C; Cacchio, M; Algeri, S. A review of specific dietary antioxidants and the effects on biochemical mechanisms related to neurodegenerative processes. *Neurobiol Aging,* 2002 23 (5):719-735.

[20] Ji, BT; Chow, WH; Hsing, AW; McLaughlin, JK; Dai, Q; Gao, YT; Blot, WJ; Fraumeni, JF. Green tea consumption and the risk of pancreatic and colorectal cancers. *Int J Cancer,* 1997 70 (3):255-258.

[21] van Dam, RM; Hu, FB. Coffee consumption and risk of type 2 diabetes: a systematic review. *JAMA,* 2005 294 (1):97-104.

[22] Khurana, S; Venkataraman, K; Hollingsworth, A; Piche, M; Tai, TC. Polyphenols: Benefits to the cardiovascular system in health and in aging. *Nutrients*, 2013 5(10):3779-3827.

[23] Wang, LS; Kuo, CT; Cho, SJ; Seguin, C; Siddiqui, J; Stoner, K; Weng, YI; Huang, TH; Tichelaar, J; Yearsley, M; Stoner, GD; Huang, YW. Black raspberry-derived anthocyanins demethylate tumor suppressor genes through the inhibition of DNMT1 and DNMT3B in colon cancer cells. *Nutr Cancer,* 2013 65 (1):118-125.

[24] Tsao, R. Chemistry and biochemistry of dietary polyphenols. *Nutrients,* 2010 2 (12):1231-1246.

[25] Kong, JM; Chia, LS; Goh, NK; Chia, TF; Brouillard, R. Analysis and biological activities of anthocyanins. *Phytochemistry,* 2003 64 (5):923-933

[26] Cao, G; Prior, RL. Anthocyanins are detected in human plasma after oral administration of an elderberry extract. *Clinical Chemistry,* 1999 45:574-576.

[27] Del Rio, D; Rodriguez-Mateos, A; Spencer, JP; Tognolini, M; Borges, G; Crozier, A. Dietary (poly)phenolics in human health: structures, bioavailability, and evidence of protective effects against chronic diseases. *Antioxid Redox Signal*, 2013 18 (14):1818-1892.

[28] Devore, EE; Kang, JH; Breteler, MM; Grodstein, F. Dietary intakes of berries and flavonoids in relation to cognitive decline. *Ann Neurol,* 2012 72 (1):135-143.

[29] Tsuda, T. Dietary anthocyanin-rich plants: biochemical basis and recent progress in health benefits studies. *Mol Nutr Food Res,* 2012 56 (1):159-170.

[30] Tanaka, Y; Sasaki, N; Ohmiya, A. Biosynthesis of plant pigments: anthocyanins, betalains and carotenoids. *Plant J,* 2008 54 (4):733-749.

[31] He, J; Giusti, MM. Anthocyanins: natural colorants with health-promoting properties. *Annu Rev Food Sci Technol,* 2010 1:163-187.

[32] Wallace, TC. Anthocyanins in cardiovascular disease. *Adv Nutr*, 2011 2 (1):1-7.

[33] Valenti, L; Riso, P; Mazzocchi, A; Porrini, M; Fargion, S; Agostoni, C. Dietary anthocyanins as nutritional therapy for nonalcoholic fatty liver disease. *Oxid Med Cell Longev,* 2013 2013:145421.

[34] Milbury, PE; Kalt, W. Xenobiotic metabolism and berry flavonoid transport across the blood-brain barrier. *J Agric Food Chem*, 2010 58 (7):3950-3956.

[35] Andres-Lacueva, C; Shukitt-Hale, B; Galli, RL; Jauregui, O; Lamuela-Raventos, RM; Joseph, JA. Anthocyanins in aged blueberry-fed rats are found centrally and may enhance memory. *Nutr Neurosci,* 2005 8 (2):111-120.

[36] Rashid, K; Wachira, FN; Nyabuga, JN; Wanyonyi, B; Murilla, G; Isaac, AO. Kenyan purple tea anthocyanins ability to cross the blood brain barrier and reinforce brain antioxidant capacity in mice. *Nutr Neurosci,* 2014 17(4):178-185.

[37] Talavéra, S; Felgines, C; Texier, O; Besson, C; Gil-Izquierdo, A; Lamaison, JL; Rémésy, C. Anthocyanin metabolism in rats and their distribution to digestive area, kidney, and brain. *J Agric Food Chem,* 2005 53 (10):3902-3908.

[38] Kalt, W; Blumberg, JB; McDonald, JE; Vinqvist-Tymchuk, MR; Fillmore, SA; Graf, BA; O'Leary, JM; Milbury, PE. Identification of anthocyanins in the liver, eye, and brain of blueberry-fed pigs. *J Agric Food Chem,* 2008 56 (3):705-712.

[39] Halliwell, B; Gutteridge, JMC. Free radicals in biology and medicine. 4. Oxford: *Clarendon,* 2007.

[40] Williams, CM; El Mohsen, MA; Vauzour, D; Rendeiro, C; Butler, LT; Ellis, JA; Whiteman, M; Spencer, JP. Blueberry-induced changes in spatial working memory correlate with changes in hippocampal CREB phosphorylation and brain-derived neurotrophic factor (BDNF) levels. *Free Radic Biol Med,* 2008 45 (3):295-305.

[41] Rendeiro, C; Vauzour, D; Kean, RJ; Butler, LT; Rattray, M; Spencer, JP; Williams CM. Blueberry supplementation induces spatial memory improvements and region-specific regulation of hippocampal BDNF mRNA expression in young rats. *Psychopharmacology* (Berl), 2012 223 (3):319-330.

[42] Lee, SG; Kim, B; Yang, Y; Pham, TX; Park, YK; Manatou, J; Koo, SI; Chun, OK; Lee, JY. Berry anthocyanins suppress the expression and secretion of proinflammatory mediators in macrophages by inhibiting nuclear translocation of NF-κB independent of NRF2-mediated mechanism. *J Nutr Biochem,* 2014 25 (4):404-411.

[43] Wang, H; Cao, G; Prior, RL. Oxygen radical absorbing capacity of anthocyanins. *J. Agric. Food Chem,* 1997 45:304-309.

[44] Mazza, G; Kay, CD; Cottrell, T; Holub, BJ. Absorption of anthocyanins from blueberries and serum antioxidant status in human subjects. *J Agric Food Chem,* 2002 50 (26):7731-7737.

[45] Wang, J; Mazza, G. Effects of anthocyanins and other phenolic compounds on the production of tumor necrosis factor alpha in LPS/IFN-gamma-activated RAW 264.7 macrophages. *J Agric Food Chem,* 2002 50 (15):4183-4189.

[46] Cisowska, A; Wojnicz, D; Hendrich, AB. Anthocyanins as antimicrobial agents of natural plant origin. *Nat Prod Commun,* 2011 6 (1):149-156.

[47] Kamei, H; Kojima, T; Hasegawa, M; Koide, T; Umeda, T; Yukawa, T; Terabe, K. Supression of tumor-cell growth by anthocyanins in vitro. *Cancer Invest,* 1995 13(6): 590-594.

[48] Katsube, N; Iwashita, K; Tsushida, T; Yamaki, K; Kobori, M. Induction of apoptosis in cancer cells by Bilberry (Vaccinium myrtillus) and the anthocyanins. *J Agric Food Chem,* 2003 51 (1):68-75.

[49] Renaud, S; de Lorgeril, M. Wine, alcohol, platelets, and the French paradox for coronary heart disease. *Lancet,* 1992 339 (8808):1523-1526.

[50] Demrow, HS; Slane, PR; Folts, JD. Administration of wine and grape juice inhibits in vivo platelet activity and thrombosis in stenosed canine coronary arteries. *Circulation,* 1995 91 (4):1182-1188.

[51] Martin, S; Giannone, G; Andriantsitohaina, R; Martinez, MC. Delphinidin, an active compound of red wine, inhibits endothelial cell apoptosis via nitric oxide pathway and regulation of calcium homeostasis. *Br J Pharmacol,* 2003 139 (6):1095-1102.

[52] Mazza, GJ. Anthocyanins and heart health. *Ann Ist Super Sanita,* 2007 43 (4):369-374.

[53] Lu, J; Wu, DM; Zheng, YL; Hu, B; Zhang, ZF. Purple sweet potato color alleviates D-galactose-induced brain aging in old mice by promoting survival of neurons via PI3K pathway and inhibiting cytochrome C-mediated apoptosis. *Brain Pathol,* 2010 20 (3):598-612.

[54] Dreiseitel, A; Korte, G; Schreier, P; Oehme, A; Locher, S; Domani, M; Hajak, G; Sand, PG. Berry anthocyanins and their aglycons inhibit monoamine oxidases A and B. *Pharmacol Res,* 2009 59 (5):306-311.

[55] Fernandez, HH; Chen, JJ. Monoamine oxidase-B inhibition in the treatment of Parkinson's disease. *Pharmacotherapy,* 2007 27 (12 Pt 2):174S-185S.

[56] Maeda-Yamamoto, M; Saito, T; Nesumi, A; Tokuda, Y; Ema, K; Honma, D; Ogino, A; Monobe, M; Murakami, A; Tachibana, H. Chemical analysis and acetylcholinesterase inhibitory effect of anthocyanin-rich red leaf tea (cv. Sunrouge). *J Sci Food Agric,* 2012 92 (11):2379-2386.

[57] Moreira, PI; Zhu, X; Nunomura, A; Smith, MA; Perry, G. Therapeutic options in Alzheimer's disease. *Expert Rev Neurother,* 2006 6 (6):897-910.

[58] Qin, L; Zhang, J; Qin, M. Protective effect of cyanidin 3-O-glucoside on beta-amyloid peptide-induced cognitive impairment in rats. *Neurosci Lett,* 2013 534:285-288.

[59] Shih, PH; Chan, YC; Liao, JW; Wang, MF; Yen, GC. Antioxidant and cognitive promotion effects of anthocyanin-rich mulberry (*Morus atropurpurea* L.) on senescence-accelerated mice and prevention of Alzheimer's disease. *J Nutr Biochem,* 2010 21 (7):598-605.

[60] Vepsäläinen, S; Koivisto, H; Pekkarinen, E; Mäkinen, P; Dobson, G; McDougall, GJ; Stewart, D; Haapasalo, A; Karjalainen, RO; Tanila, H; Hiltunen, M. Anthocyanin-enriched bilberry and blackcurrant extracts modulate amyloid precursor protein processing and alleviate behavioral abnormalities in the APP/PS1 mouse model of Alzheimer's disease. *J Nutr Biochem,* 2013 24 (1):360-370.

[61] Kim, HG; Ju, MS; Shim, JS; Kim, MC; Lee, SH; Huh, Y; Kim, SY; Oh, MS. Mulberry fruit protects dopaminergic neurons in toxin-induced Parkinson's disease models. *Br J Nutr,* 2010 104 (1):8-16.

[62] Moreira, EL; Rial, D; Duarte, FS; de Carvalho, CR; Horst, H; Pizzolatti, MG; Prediger, RD; Ribeiro-do-Valle, RM. Central nervous system activity of the proanthocyanidin-rich fraction obtained from *Croton celtidifolius* in rats. *J Pharm Pharmacol,* 2010 62 (8):1061-1068.

[63] Moreira, EL; Rial, D; Aguiar, AS; Figueiredo, CP; Siqueira, JM; DalBó, S; Horst, H; de Oliveira, J; Mancini, G; dos Santos, TS; Villarinho, JG; Pinheiro, FV; Marino-Neto, J; Ferreira, J; De Bem, AF; Latini, A; Pizzolatti, MG; Ribeiro-do-Valle, RM; Prediger, RD. Proanthocyanidin-rich fraction from *Croton celtidifolius* Baill confers neuroprotection in the intranasal 1-methyl-4-phenyl-1,2,3,6-tetrahydropyridine rat model of Parkinson's disease. *J Neural Transm,* 2010 117 (12):1337-1351.

[64] Assis, LC; Hort, MA; de Souza, GV; Martini, AC; Forner, S; Martins, DF; Silva, JC; Horst, H; Dos Santos, AR; Pizzolatti, MG; Rae, GA; Koepp, J; de Bem, AF; do Valle, RM. Neuroprotective effect of the

proanthocyanidin-rich fraction in experimental model of spinal cord injury. *J Pharm Pharmacol,* 2014 66(5):694-704.

[65] Strathearn, KE; Yousef, GG; Grace, MH; Roy, SL; Tambe, MA; Ferruzzi, MG; Wu, QL; Simon, JE; Lila, MA; Rochet, JC. Neuroprotective effects of anthocyanin- and proanthocyanidin-rich extracts in cellular models of Parkinson's disease. *Brain Res,* 2014 1555:60-77.

[66] Jennings, A; Welch, AA; Fairweather-Tait, SJ; Kay, C; Minihane, AM; Chowienczyk, P; Jiang, B; Cecelja, M; Spector, T; Macgregor, A; Cassidy, A. Higher anthocyanin intake is associated with lower arterial stiffness and central blood pressure in women. *Am J Clin Nutr,* 2012 96 (4):781-788.

[67] McCullough, ML; Peterson, JJ; Patel, R; Jacques, PF; Shah, R; Dwyer, JT. Flavonoid intake and cardiovascular disease mortality in a prospective cohort of us adults. *Am J Clin Nutr,* 2012 95:454–464.

[68] Cassidy, A; Mukamal, KJ; Liu, L; Franz, M; Eliassen, AH; Rimm, EB. High anthocyanin intake is associated with a reduced risk of myocardial infarction in young and middle-aged women. *Circulation,* 2013 127 (2):188-196.

[69] Golbe, LI; Farrell, TM; Davis, PH. Case-control study of early life dietary factors in Parkinson's disease. *Arch Neurol,* 1988 45 (12):1350-1353.

[70] Krikorian, R; Shidler, MD; Nash, TA; Kalt, W; Vinqvist-Tymchuk, MR; Shukitt-Hale, B; Joseph, JA. Blueberry supplementation improves memory in older adults. *J Agric Food Chem,* 2010 58 (7):3996-4000.

[71] Krikorian, R; Nash, TA; Shidler, MD; Shukitt-Hale, B; Joseph, JA. Concord grape juice supplementation improves memory function in older adults with mild cognitive impairment. *Br J Nutr,* 2010 103 (5):730-734.

[72] Krikorian, R; Boespflug, EL; Fleck, DE; Stein, AL; Wightman, JD; Shidler, MD; Sadat-Hossieny, S. Concord grape juice supplementation and neurocognitive function in human aging. *J Agric Food Chem,* 2012 60 (23):5736-5742.

[73] Lindsay, J; Laurin, D; Verreault, R; Hébert, R; Helliwell, B; Hill, GB; McDowell, I. Risk factors for Alzheimer's disease: a prospective analysis from the Canadian Study of Health and Aging. *Am J Epidemiol,* 2002 156 (5):445-453.

[74] Vasanthi, HR; Parameswari, RP; DeLeiris, J; Das, DK. Health benefits of wine and alcohol from neuroprotection to heart health. *Front Biosci* (Elite Ed), 2012 4:1505-1512.

[75] Orgogozo, JM; Dartigues, JF; Lafont, S; Letenneur, L; Commenges, D; Salamon, R; Renaud, S; Breteler, MB. Wine consumption and dementia in the elderly: a prospective community study in the Bordeaux area. *Rev Neurol* (Paris), 1997 153 (3):185-192.

[76] Rivero-Pérez, MD; Muñiz, P; González-Sanjosé, ML. Contribution of anthocyanin fraction to the antioxidant properties of wine. *Food Chem Toxicol,* 2008 46 (8):2815-2822.

[77] Waterhouse, AL. Wine phenolics. *Ann N Y Acad Sci,* 2002 957:21-36.

[78] Arranz, S; Chiva-Blanch, G; Valderas-Martínez, P; Medina-Remón, A; Lamuela-Raventós, RM; Estruch, R. Wine, beer, alcohol and polyphenols on cardiovascular disease and cancer. *Nutrients,* 2012 4 (7):759-781.

[79] Conlin, PR. The dietary approaches to stop hypertension (DASH) clinical trial: implications for lifestyle modifications in the treatment of hypertensive patients. *Cardiol Rev,* 1999 7 (5):284-288.

[80] van't Veer, P; Jansen, MC; Klerk, M; Kok, FJ. Frutis and vegetables in the prevention of cancer and cardiovascular disease. *Public Health Nutr,* 2000 3(1):103-107.

[81] Dontas, AS; Zerefos, NS; Panagiotakos, DB; Vlachou, C; Valis, DA. Mediterranean diet and prevention of coronary heart disease in the elderly. *Clin Interv Aging,* 2007 2 (1):109-115

[82] Johnston, C. Functional Foods as Modifiers of Cardiovascular Disease. *Am J Lifestyle Med,* 2009 3 (1 Suppl):39S-43S.

[83] Lourida, I; Soni, M; Thompson-Coon, J; Purandare, N; Lang, IA; Ukoumunne, OC; Llewellyn, DJ. Mediterranean diet, cognitive function, and dementia: a systematic review. *Epidemiology,* 2013 24 (4):479-489.

[84] Phillips, HS; Hains, JM; Armanini, M; Laramee, GR; Johnson, SA; Winslow, JW. BDNF mRNA is decreased in the hippocampus of individuals with Alzheimer's disease. *Neuron,* 1991 7 (5):695-702.

[85] Schaaf, MJ; Workel, JO; Lesscher, HM; Vreugdenhil, E; Oitzl, MS; de Kloet, ER. Correlation between hippocampal BDNF mRNA expression and memory performance in senescent rats. *Brain Res,* 2001 915 (2):227-233.

[86] Pang, PT; Teng, HK; Zaitsev, E; Woo, NT; Sakata, K; Zhen, S; Teng, KK; Yung, WH; Hempstead, BL; Lu, B. Cleavage of proBDNF by

tPA/plasmin is essential for long-term hippocampal plasticity. *Science,* 2004 306 (5695):487-491.

[87] Aguiar, AS; Castro, AA; Moreira, EL; Glaser, V; Santos, AR; Tasca, CI; Latini, A; Prediger, RD. Short bouts of mild-intensity physical exercise improve spatial learning and memory in aging rats: involvement of hippocampal plasticity via AKT, CREB and BDNF signaling. *Mech Ageing Dev,* 2011 132 (11-12):560-567.

[88] Morris, JC; Price, JL. Pathologic correlates of nondemented aging, mild cognitive impairment, and early-stage Alzheimer's disease. *J Mol Neurosci,* 2001 17:101-118.

[89] Morris, JC. Early-stage and preclinical Alzheimer disease. *Alzheimer Dis Assoc Disord,* 2005 19:163-165.

[90] Kivipelto, M; Solomon, A. Alzheimer's disease - the ways of prevention. *J Nutr Health Aging,* 2008 12 (1):89S-94S.

In: Proanthocyanidins
Editor: Ingrid Sullivan

ISBN: 978-1-63463-430-4

Chapter 5

PROANTHOCYANIDINS AND CANCER: EFFECTS OF PROANTHOCYANIDINS ON CANCER CELL BEHAVIOURS *IN VITRO*

Robert A. R. Hurta*
Department of Biology, University of Prince Edward Island,
Charlottetown , P. E., Canada

ABSTRACT

Proanthocyanidins are a class of "bioactive" compounds found in a number of natural products and can be complex in their structure and in their composition. Proanthocyanidins have been found in grape seeds, blueberries, cranberries and apples amongst many other sources. This paper reviews the existing recent research on the anti-cancer properties of proanthocyanidins with a particular focus on the proanthocyanidins found in grape seeds, blueberries, cranberries, and apples. Proanthocyanidins have been proposed to exert their anti-cancer effects by either inhibiting cellular proliferation, by the induction of apoptosis, by affecting the inflammatory response or by inhibiting the invasive capabilities of cancer cells, amongst other possible mechanisms. These proanthocyanidin – mediated effects involve alterations in a number of cellular activities and also involve alterations in various cellular signal transduction mechanisms. Collectively, these findings suggest a potential role for

* Corresponding Author: Robert Hurta, Department of Biology, University of Prince Edward Island, Charlottetown, P.E. Canada , C1A4P3. E:mail : rhurta@upei.ca

proanthocyanidins found in specific natural products to possess possible chemo-preventative and/or possible chemo-protective properties against cancer cells and as such, supports the inclusion of these proanthocyanidin-containing natural products in an "anti-cancer" diet.

INTRODUCTION

This review describes studies which have examined the possible anti-cancer effects of proanthocyanidins found in a number of fruits. Proanthocyanidins are a class of flavonoids which contribute to the phytochemical profile of a number of natural products including many fruits. Proanthocyanidins possess anti-oxidant properties. Proanthocyanidins have been suggested to possess a number of health benefits including possessing anti-cancer effects, anti-microbial effects, providing cardio-protective effects and to influence neuro-protective activities [1]. This current review paper summarizes recent research findings from studies which have examined the effects of proanthocyanidins on the behaviours of cancer cells. This current review paper will specifically focus on recent studies which examined the effects of proanthocyanidins isolated from grape seeds, cranberries, blueberries and apples and will further be restricted to presenting those effects on cancer cells *in vitro*.

Chemistry of Proanthocyanidins

Proanthocyanidins (PACs) or condensed tannins are amongst the most abundant polyphenols in our diet and may play key roles in a number of disease etiologies. The biological activity of plant proanthocyanidins depends on their chemical structure and concentration [2]. The proanthocyanidins are structurally diverse and inherently complex. Proanthocyanidins or condensed tannins are polymers of flavan-3-ol. The structure of proanthcyanidins varies according to the hydroxylation pattern of the A and B rings, the stereochemistry of the C2, C3 and C4 of the central ring and the interflavan linkages. The most common structural monomeric units of PACs in plants are (+)afzelechin, (+) catechin and (+) gallocatechin (2R:3S forms) and their diasteromers (-)epiafzelechin, (-) epicatechin and (-) epigallocatechin (2S:3S forms) all of which contain a phloroglucinol type-A ring [2]. Esterification with other molecules is also possible [2]. Proanthocyanidins that are entirely composed of (epi)catechin are called procyanidins (2). Due to their diversity and structural complexity the analysis and characterization of proanthocyanidins can be challenging.

Grape Seed Proanthocyanidins: Effects on Cancer Cell Behaviours In Vitro

Considerable research has been done on the effects of grape seed proanthocyanidins on the behaviour of cancer cells *in vitro*. The following is a discussion of some of the recent findings which illustrate the effects of grape seed proanthocyanidins on various cancer cell lines *in vitro* with a focus on the mechanism of action and the signal transduction pathways involved.

Oligomeric proanthocyanidin complexes (OPC) extracted from grape seeds have been found to exert anti-proliferative and pro-apoptotic effects on prostate cancer cells (3). In this regard, androgen sensitive LNCaP cells are more responsive to OPC in terms of inhibition in comparison to androgen insensitive or refractory PC3 or DU145 cells. Treatment with OPC resulted in a decrease in the number of LNCaP cells in the S phase and an increase in the percentage of cells in sub G1 phase (3). The anti-proliferative effect of OPCs in the LNCaP cell line was associated with down-regulation of the androgen receptor [3]. Additionally, inhibition of the expression of cyclin dependent kinases and cyclins and the stimulation of tumor suppressors p21 and p27 were seen in LNCaP and PC3 cells after treatment with OPC [3]. Alteration in the Bcl-2/Bax ratio was also observed in LNCaP and PC3 cells after treatment with OPC [3]. Such alterations would favour the induction of apoptosis or programmed cell death in these cells. OPC caused an increase of phosphorylated mitogen activated protein kinase p44 and p42 proteins which was suggestive of an induction of cellular differentiation in response to OPC treatment [3].

The inhibition of expression of matrix metalloproteinases (MMPs) in human prostate carcinoma cells in response to grape seed proanthocyanidins has also been demonstrated [4]. Matrix metalloproteinases are key cellular activities associated with the invasion and metastasis of cancer cells. Grape seed proanthocyanidin-mediated inhibition of MMPs in these prostate cancer cells was associated with inhibition of the activation of mitogen activated protein kinase (MAPK) and nuclear factor kappa B (NFκB) [4]. In DU145 androgen insensitive human prostate cancer cells, grape seed proanthocyanidins caused a dose dependent inhibition of cell growth and proliferation [4]. Additionally, the grape seed proanthocyanidins attenuated the fibroblast conditioned medium- induced expression of MMP-2 and MMP-9 in these DU145 human prostate cancer cells [4]. Associated with this behaviour, an inhibition of fibroblast conditioned medium- induced phosphorylation of extracellular signal regulated kinase (ERK 1/2) and p38 was noted however the grape seed proanthocyanidin had little effect on c-Jun N-terminal kinase under similar conditions [4]. In response to treatment of DU145 human prostate cancer cells with grape seed proanthocyanidin extract, a dose dependent inhibition of the fibroblast condition medium –induction of NFκB occurred in these DU145 human prostate cancer cells *in vitro* [4]. This inhibition of NFκB induction occurred in conjunction with inhibition of both MMP-2 and MMP-9. Collectively, these findings suggest that the inhibition of these particular MMPs (which are suggested to be metastasis related activities in tumor cells) by grape seed proanthocyanidins is linked with the inhibition of activation of MAPK and NFκB pathways [4].

Grape seed proanthocyanidins also inhibit the growth of cervical cancer cells by inducing apoptosis mediated by the mitochondrial pathway [5]. Apoptosis or programmed cell death is a highly regulated and ordered form of cellular death. *In vitro* treatment of HeLa and SiHa cells with grape seed proanthocyanidins resulted in a pronounced inhibition of cell viability which was associated with a dose dependent induction of apoptosis [5]. Aspects of the underlying mechanisms associated with this apoptosis were elucidated. The grape seed proanthocyanidin mediated- induction of apoptosis within these cervical cancer cells was associated with increased expression of the pro-apoptotic protein Bak-1 coupled with decreased expression of the anti-apoptotic protein Bcl-2 [5]. Furthermore, the loss of mitochondrial membrane potential and the activation of caspase-3 which occurred in response to grape seed proanthocyanidin treatment suggested that the grape seed proanthcyanidin induction of apoptosis in these cervical cancer cells occurred via the mitochondrial pathway of apoptosis [5].

Grape seed proanthocyanidins can also induce apoptosis and inhibit metastasis of highly metastatic breast carcinoma cells [6]. In these studies, the *in vitro* treatment of breast cancer cells 4T1 and MCF-7 with grape seed proanthocyanidins resulted in a marked inhibition of cellular proliferation and viability. Additionally, an induction of apoptosis in 4T1 cells occurred in response to grape seed proanthocyanidins in a time and dose dependent manner [6]. Alterations in the ratio of Bax/Bcl-2 proteins in favour of promoting apoptosis was documented. This induction of apoptosis was linked to the release of cytochrome C, increased Apaf-1, activation of caspase-3 and poly(ADP-ribose) polymerase [6]. Following treatment of 4T1 cells with Z-VAD-FMK (which is a pan-caspase inhibitor), a partial but significant inhibition of apoptosis in 4T1 cells resulted which was suggestive of an involvement of both caspase activation dependent and activation independent pathways of apoptosis of 4T1 cells which occurs in response to grape seed proanthocyanidins [6].

Grape seed proanthocyanidins can also suppress breast cell carcinogenesis induced by chronic exposure to combined 4-(methylnitroamino)-1-(3-pyridyl)-1 butanone and benzo[a]pyrene (two carcinogens) treatments [7]. This suppression of breast cell carcinogenesis was linked to grape seed proanthocyanidin's ability to diminish the expression of cytochrome P-450 enzymes (specifically, CYP1A1 and CYP1B1) which can bio-activate these two carcinogens [7]. This reduction in expression of these cytochrome P-450 enzymes was suggested to possibly contribute to the preventative mechanism of grape seed proanthocyanidin extract in suppression of pre-cancerous cellular carcinogenesis. It was proposed that these findings suggest that grape seed proanthocyanidin extract may have a value in the prevention of breast cancer carcinogenesis induced by repeated exposures to low doses of multiple environmental carcinogens [7].

Interestingly, proanthocyanidins from grape seeds inactivate the P-I-3 kinase/PKB pathway and induce apoptosis in colon cancer cell lines [8]. Grape seed proanthocyanidin extract significantly inhibited cellular viability and increased apoptosis in CaCo-2 cells but did not alter the cellular viability in the normal colon cell line (NCM460) [8] suggesting that the grape seed proanthocyanidin extract has the ability to selectively inhibit only colon cancer cells with no apparent effects on normal or untransformed colon cells. The increased apoptosis in CaCo-2 cells in response to grape seed proanthocyanidin extract treatment correlated with an attenuation of P-I-3 kinase (p110 and p85 subunits) and decreased protein kinase B [Ser(473)] phosphorylation [8]. The beneficial effects of grape seed proanthocyanidin extract against colon cancer cells occurs then by means of increased apoptosis and the suppression of the important P-I-3 kinase survival -related pathway. Furthermore, grape seed proanthocyanidin extract can also inhibit colon cancer induced angiogenesis (the process whereby a blood supply is acquired) through suppression of the expression of vascular endothelial growth factor (VEGF) and angiopoietin -1(Ang1) [9]. The mechanism of action of the grape seed proanthocyanidins was related to the inhibition of expression of VEGF and Ang-1. This inhibition mechanism was proposed to also involve scavenging reactive oxygen species and this was further suggested as a possible mechanism of action of the grape seed proanthocyanidins [9]. As such, grape seed proanthocyanidin extract's ability to inhibit VEGF and Ang-1 expression thereby affecting angiogenesis coupled with affecting reactive oxygen species expression collectively can account for grape seed proanthocyanidin extract's ability to inhibit colon cancer cell growth. In another study, grape seed proanthocyanidins in combination with eicosapentaenoic acid-free fatty acid, and epigallocatechin -3-gallate had a marked effect on mTOR signalling in colorectal cancer cells [10]. The mammalian target of rapamycin (mTOR) plays a critical role in colorectal cancer by regulating protein translation and controlling cell growth , proliferation, metabolism and survival. These three agents collectively completely inhibited the mTOR signaling pathway in HCT116 and SW480 colon cancer cells [10]. The altered biology of the colon cancer cells in response to this particular treatment was associated with changes in protein translation of ribosomal proteins, c-Myc and cyclin D1 and a reduction of clonal capability in both cell lines with a block of the cell cycle in G_0-G_1.[10]. The induction of apoptosis also occurred [10].

Cell migration is a fundamental aspect of cancer cells' metastasis. The urokinase plasminogen activator system (uPA) is implicated in cancer cell migration and cancer cell metastasis [11]. The efficacy of grape seed proanthocyanidins in regulating uPA expression and cell migration using highly metastatic androgen independent PC3 prostate cancer cells as a model was determined [12]. Invasion assays revealed the inhibitory effect of grape seed proanthocyanidin extract on PC3 cell migration. Grape seed proanthocyanidin extract down- regulated uPA. This down-regulation response was extract concentration dependent [12]. From a cellular signalling perspective, grape seed proanthocyanidin extract inhibited the DNA binding activity of NFκB which then resulted in decreased NFκ B-dependent- uPA transcription [12].

Grape seed proanthocyanidins can also inhibit pancreatic cancer cell growth *in vitro* (and i*n vivo*). The grape seed proanthocyanidin extract mediated inhibition of pancreatic cancer cell growth was shown to involve the induction of apoptosis and to involve alterations in the P-I-3K/Akt signalling pathway [13]. Specifically, treatment of human pancreatic cancer cells (MIA Paca-2, PANC-1 and AsPC-1) with grape seed proanthocyanidins *in vitro* reduced cell viability and increased G2/M phase arrest of the cell cycle leading to induction of apoptosis in a dose and time dependent manner [13]. In agreement with this study, grape seed procyanidins also can inhibit cell growth and increase apoptosis in pancreatic cancer cells [14]. Grape seed procyanidins extract inhibits cell proliferation and increases apoptosis in pancreatic adenocarcinoma MIA PaCa-2 cells. This grape seed procyanidin- mediated induction of apoptosis was primarily mediated by the down-regulation of the anti-apoptotic protein Bcl-2 and the depolarization of the mitochondrial membrane [14]. The grape seed procyanidins also reduced the formation of reactive oxygen species. The component of the extract that possesses the highest anti-proliferative and pro-apoptotic activity was determined to be gallic acid [14]. Other studies have shown that grape seed proanthocyanidins can inhibit the migration potential of pancreatic cancer cells by affecting the mesenchymal to epithelial transition and targeting NFκB [15]. Inhibition of cell migration by grape seed proanthocyanidins was associated with the reversal of the epithelial-to mesenchymal transition which was associated with the up-regulation of E-cadherin and desmoglein-2 expression and the down-regulation of fibronectin, N-cadherin and vimentin expression [15].

Grape seed proanthocyanidins can also affect the cellular behaviour and biology of non small cell lung cancer cells *in vitro*. The inhibition of non small cell lung cancer cell migration by grape seed proanthocyanidins has been shown to be mediated through the inhibition of nitric oxide , guanylate cyclase and ERK 1/2 [16]. When A549 and H1299 cells (non small cell lung cancer cell lines) were exposed to grape seed proanthocyanidins a concentration dependent inhibition of cellular migration occurred [16]. Elevated levels of endogenous nitric oxide and nitric oxide synthases found in A549 and H1299 cells were suppressed by grape seed proanthocyanidins [16]. Furthermore, grape seed proanthocyanidins were shown to down-regulate the migration promoting capacity of L-arginine [16]. Grape seed proanthocyanidins also can reduce elevated levels of cGMP in cancer cells and also blocked the migration restoring activity of 8-Br-cGMP [16]. Grape seed proanthocyanidins can also inhibit the *in vitro* (and *in vivo*) growth of human non small cell lung cancer cells by inhibiting prostaglandin E(2) and prostaglandin E(2) receptor expression [17]. Grape seed proanthocyanidins can specifically modulate the over-expression of cyclooxygenase-2 (COX-2) and PGE(2) and the PGE(2) receptors (EP1 and EP4) [17]. Grape seed proanthocyanidins can also inhibit cigarette smoke condensate- induced lung cancer cell migration through inhibition of NADPH oxidase and reduction in the binding of p22phox and p47 phox proteins [18]. Grape seed proanthocyanidins capacity to induce apoptosis in non small lung cancer cells occurs via the loss of mitochondrial membrane potential of human non small lung cancer cells *in vitro* (and *in vivo*) [19]. Associated with this induction of apoptosis was an increased expression of pro-apoptotic proteins, a decreased expression of anti-apoptotic proteins, disruption of mitochondrial membrane potential and activation of caspases 9 and 3 and of poly (ADP-ribose) polymerase (PARP) [19]. Additionally, grape seed proanthocyanidins can also have effects on the cell cycle and cell cycle associated activities within non small lung cancer cells. Treatment of A549 and H1299 cells resulted in an increase in G1 arrest, coupled with a marked increased expression of cyclin dependent kinase inhibitors (Cip1/p21 and Kip1/p27)[19]. A simultaneous decrease in the level of Cdk2, Cdk4, Cdk6 and cyclins occurred in A549 and H1299 non small lung cancer cells following treatment of these cells with grape seed proanthocyanidins [19].

Grape seed proanthocyanidins also possess the ability to reactivate silenced tumor suppressor genes in human skin cancer cells by targeting epigenetic regulators [20]. Treatment of A431 and SCC13 human squamous cell carcinoma cell lines with grape seed proanthocyanidins resulted in decreased levels of global DNA methylation, 5-methylcytosine, DNA methyltransferase activity and mRNA and protein levels of DNMT1, DNMT3a and DNMT3b in these cells [20]. Grape seed proanthocyanidins also decreased histone deacetylase activity, increased levels of acetylated lysines 9 and 14 on histone H3 and acetylated lysines 5, 12, and 16 on histone H4 and reduced the levels of methylated H3-lysine 9 [20]. Grape seed proanthocyanidin treatment also resulted in the re-expression of the mRNA and proteins of silenced tumor suppressor genes, RASSF1A, p16(INK4a) and Cip1/p21 [20]. These findings illustrate aspects of the epigenetic mechanisms whereby grape seed proanthocyanidins can modulate cancer cell behaviour *in vitro*.

In summary, grape seed proanthocyanidins can effectively modulate the behaviour of a number of cancer cells i*n vitro* which results in the inhibition of cancer cell growth. This modulation of cancer cell behaviour *in vitro* by grape seed proanthocyanidins occurs through a number of regulatory mechanisms which include alterations in key cellular signal transduction pathways.

Cranberry and Blueberry Proanthocyanidins: Effects on Cancer Cell Behaviours In Vitro

Both cranberries and blueberries are rich in proanthocyanidins (PAC). Cranberries have been evaluated to have the highest average PAC content amongst a variety of fruits tested [21]. Blueberry PACs content is also high with lowbush blueberries higher in proanthocyanidin content than high bush blueberry [21]. A number of research investigations have illustrated the ability of cranberry and blueberry proanthocyanidins to modulate the cell growth behaviour of cancer cells *in vitro*. In this regard, a whole cranberry extract containing PACs (and other flavonoids) inhibited ornithine decarboxylase (ODC) activity in mouse epithelial cells [22]. ODC is a key rate limiting and regulatory activity involved in polyamine biosynthesis and ODC activity is intimately linked to cell growth [23]. A proanthocyanidin fraction isolated from whole cranberry fruit was observed to selectively inhibit the growth of H460 human large cell lung carcinoma, HT-29 colon adenocarcinoma, and K562 chronic myelogenous leukemia cells in an eight tumor cell line panel [21]. Inhibition of acid induced proliferation of SEG-1 esophageal cells by cranberry PACs was shown to be accompanied by cell cycle arrest at the G1 checkpoint coupled with morphological changes characteristic of apoptosis [24]. Cranberry proanthocyanidins (and ursolic acid) induce apoptosis in HCT116 and HT29 colon carcinoma cells [25]. Cranberry proanthocyanidins down-regulate the expression of the inhibitors of apoptosis proteins (IAPs) and other anti-apoptotic molecules in the H460 lung cancer cell line [26] consistent with earlier studies that found reduced cell proliferation of the cell line with cranberry proanthocyanidin fractions [27].

Matrix metalloproteinase are proteins intimately linked to tumour cell invasion and metastasis formation. Increased expression of matrix metalloproteinases (MMPs) in particular, matrix metalloproteinase-2 and matrix metalloproteinase-9 expression is often associated with enhanced tumour cell growth and increased invasive potential [28]. Treatment of DU145 human prostate cancer cells with proanthocyanidins isolated from whole cranberry fruit reduced the activity of matrix metalloproteinase -2 (MMP-2) and matrix metalloproteinase-9 (MMP-9) significantly in as little as 1 hour post treatment with cranberry proanthocyanidins and this inhibition of MMP activity occurred independent of cytotoxicity which occurred after 6 hours of treatment [29]. This observation suggested that the cranberry proanthocyanidins' effect on MMP activity was target –directed and not just an aspect of a generalized cellular cytotoxicity effect. The inhibition of MMP-2 and MMP-9 in response to cranberry proanthocyanidin which occurred in DU145 human prostate cancer cells was accompanied by increased expression of tissue inhibitor of matrix metalloproteinase -2 (TIMP-2) and a decreased expression of the MMP inducer, EMMPRIN (extracellular MMP inducer) (29). From a cellular signalling perspective, cranberry proanthocyanidin treatment of DU145 human prostate cancer cells resulted in decreased expression of P-I-3kinase and AKT pathway associated proteins and increased phosphorylation of both p38 and ERK 1/2 [29]. Additionally, cranberry proanthocyanidins also decreased the translocation of the NFκB p65 protein to the nucleus [29]. Cranberry proanthocyanidins increased c-jun and decreased c-fos protein levels in DU145 human prostate cancer cells [29].

Collectively, these research findings suggest that cranberry proanthocyanidins inhibit matrix metalloproteinase activity in DU145 human prostate cancer cells *in vitro* via multiple regulatory and signalling mechanisms which include the induction and/or inhibition of specific temporal matrix metalloproteinase regulators and by affecting the phosphorylation status and/or expression of the mitogen activated protein (MAP) kinase, P-I-3 kinase , NFkB and AP-1 pathway proteins [29]

A flavonoid rich extract of lowbush blueberry (*Vaccinium angustifolium*) inhibited expression of MMP-2 and MMP-9 in DU145 human prostate cancer cells *in vitro* [30]. This inhibition of matrix metalloproteinase activity was attributed to the proanthocyanidins based on the activity of the crude and anthocyanin enriched fractions [30]. Treatment with the blueberry PACs also increased expression of TIMP-2 [31]. The PACs mediated inhibition of MMP-2 and MMP-9 involved possible activation of the protein kinase –C and mitogen activated protein kinase pathways but was apparently independent of the P-I-3kinase pathway [31]. Wild blueberry proanthocyanidin dimers can also affect both the promotion and proliferation of stages of carcinogenesis [32]. Other studies examining the effect of blueberry proanthocyanidin anti-proliferative effects on LNCaP and DU145 human prostate cancer cells which are androgen-sensitive and androgen –insensitive or refractory cells , respectively, may involve a number of mechanisms including effects on modulation of the androgen receptor behaviours and modulation of growth factor production and receptor expression [33].

Proanthocyanidins from the American cranberry (*Vaccinium macrocarpon*) can also induce cell cycle alterations in DU145 human prostate cancer cells *in vitro* by affecting the expression of cell cycle associated proteins [34]. In this regard, proanthocyanidin treatment of DU145 cells resulted in an increased proportion of cells in the G2-M phase of the cell cycle and a decreased proportion of cells in the G1 phase of the cell cycle [34]. These alterations in cell cycle parameters were associated with changes in the protein expression of specific cell cycle regulatory proteins and other cell cycle associated proteins. Cranberry proanthocyanidins increased the protein expression levels of cyclin E, cyclin D1, CDK2 and CDK4 and decreased the protein expression levels of cyclin A and cyclin B1 [34]. Within these DU145 human prostate cancer cells, the protein expression level of p27 increased, and the protein expression levels of $p16^{INK4}$, p21, and pRBp107 decreased in response to treatment with cranberry proanthocyanidins , whereas the protein expression level of pRBp130 was apparently unchanged [34].

Proanthocyanidins from the American cranberry (*Vaccinium macrocarpon)* have also been shown to induce programmed cell death or apoptosis in DU145 human prostate cancer cells *in vitro* and this was shown to occur via the intrinsic pathway of apoptosis [35]. Cranberry proanthocyanidin treatment of DU145 human prostate cancer cells *in vitro* led to the inhibition of the receptors for the extrinsic pathway thereby ensuring the cranberry proanthocyanidin fraction induced apoptosis occurred exclusively through the intrinsic pathway [35]. Cranberry proanthocyanidin treatment of DU145 human prostate cancer cells *in vitro* resulted in an inhibition of the protein expression levels of Fas, TRAIL, FADD, Bax, and Bcl-2 and an induction of Bid cleavage and the activation of t-Bid with the subsequent induction of the protein expression level of cytochrome –C [35]. Cranberry proanthocyanidins also inhibited the mitochondrial release of survivin, Smac/DIABLO and induced the protein expression level of PAR-4 (prostate apoptosis response-4 protein) and the caspase-3 mediated cleavage of PARP [35]. Interestingly, PAR-4 (which is a pro-apoptotic protein) can induce apoptosis independently from the activation of either the extrinsic or intrinsic pathways. Activation of PAR-4 leads to cleavage of pro-caspase-8 and since caspase -8 cleaves Bid to t-Bid, the induction of PAR-4 protein expression by cranberry proanthocyanidins may be the mechanism whereby the proanthocyanidins - mediated induction of Bid cleavage occurs [35]. The *in vitro* cranberry proanthocyanidin treatment of DU145 human prostate cancer cells induced apoptosis by up-regulating the protein expression levels of intrinsic pro-apoptotic proteins culminating in the induction of the final stages of apoptosis, poly-ADP-ribose protein (PARP) cleavage and nucleosome formation.

MicroRNA (small noncoding endogenous RNA) alterations in Barrett' s esophagus, esophageal adenocarcinoma, and esophageal adenocarcinoma cell lines occured following treatment with cranberry extracts [36]. The ability of a cranberry proanthocyanidin- rich extract to modulate miRNA expression patterns in three human EAC cell lines (JHEso-Ad-1, OE33 and OE19) was examined and this proanthocyanidin- enriched extract was able to modify miRNA profiles in these cells [36].

The anti-angiogenic activity of cranberry proanthocyanidins and their cytotoxic properties in ovarian cancer cells has been examined [37]. Proanthocyanidins from cranberry were found to express a dual activity possessing the ability to target both ovarian cancer (SKOV-3) cells' viability and angiogenesis *in vitro* [37]. These cranberry proanthocyanidins reduced the viability of both SKOV-3 and HUVEC endothelial cells and this occurred in a dose dependent manner and this was accompanied by a proanthocyanidin mediated blockage of the activation of AKT (a pro-survival factor) [37]. Additionally, cranberry proanthocyanidins inhibited vascular endothelial growth factor (VEGF) stimulated receptor phosphorylation in endothelial cells and this was correlated with the inhibition of endothelial tube formation *in vitro* [37]. Cranberry proanthocyanidins have also been shown to sensitize SKOV-3 platinum resistant ovarian cancer cells to paraplatin [38].

As evident with grape seed proanthocyanidins, cranberry and blueberry proanthocyanidins also possess the ability to modulate the behaviour of a number of cancer cells *in vitro*. The ability of cranberry and blueberry proanthocyanidins to modulate cancer cell behaviour *in vitro* also involves effects on key cellular regulatory activities and also involves alterations in a number of temporal regulators and alterations in key cellular signaling pathways.

Apple Proanthocyanidins: Effects on Cancer Cell Behaviours In Vitro

Proanthocyanidins isolated from apples are also capable of affecting the behaviour of cancer cells *in vitro*. In this regard, apple procyanidins have been shown to induce tumor cell apoptosis through the mitochondrial pathway mediated activation of caspase -3 in B16 melanoma cells and in BALB-MCE 12 mouse mammary tumor cells [39]. The apple procyanidins increased mitochondrial membrane permeability and cytochrome-C release from mitochondria and activated caspase-3 and caspase -9 within these tumor cells (39). Additionally, the induction of DNA fragmentation and apoptosis in the human stomach cancer cell line, KATO III in response to B- and C-type procyanidins from apple has been shown [40]. Apple procyanidins also have chemo-preventative properties and can affect intracellular signalling pathways and trigger apoptosis in a human adenocarcinoma derived metastatic cell line , SW620 [4]. The inhibition of SW640 cell growth in response to treatment with apple procyanidins was accompanied by inhibition of protein kinase C activity and a significant increase in extracellular signal regulated kinase 1/ 2 activity and c-jun N-terminal kinase expression with a down-regulation of polyamine biosynthesis and the enhancement of apoptosis via the activation of caspase -3 [42]. Additionally, apple procyandins possess the ability to diminish the activities of ornithine decarboxylase (ODC) and S-adenosyl methionine decarboxylase (SAMDC) and induce spermine spermidine N1-acetyl transferase (SSAT) activity with a consequent decrease in cellular polyamine levels and an accumulation of N(1)-acetyl polyamines [41].

This apple procyanidins- mediated enhancement of polyamine catabolism and consequent reduction in polyamine biosynthesis activity was similar to the effects which occurred in response to inducers of SSAT. Interestingly, the apple procyanidin- mediated effects occurred in the absence of the cytotoxicity effects associated with these SSAT inducers [41].

Apple procyanidins can also affect expression of the ErbB receptor tyrosine kinase family *in vitro* [43]. Fractions obtained from apple juice extract (≥86% oligomeric procyanidins) can also suppress epidermal growth factor receptor (EGFR) activity in cells with ErbB3 receptor activity most potently decreased [43]. The preceding research findings all suggest that apple procyanidins (proanthocyanidins) can effectively modulate the expression of cancer cells *in vitro.*

CONCLUSION

Proanthocyanidins from grape seeds, cranberry, blueberry and apple all possess the ability to alter the behaviour of several cancer cells *in vitro.* Although a number of diverse effects of these proanthocyanidins on cancer cell behaviours have been noted, there are some recurring themes whereby these proanthocyanidins act.

The inhibition of proliferation –linked activities, the induction of apoptosis, alterations in cell cycle and cell cycle-related activities, inhibition of invasion –linked activities and inhibition of angiogenesis are some of the key mechanisms which proanthocyanidins employ in order affect cancer cell growth-associated activities *in vitro.* Proanthocyanidins from grape seeds, cranberry, blueberry and apple (amongst other sources) can regulate such behaviours of cancer cells *in vitro* and this is accomplished by affecting the expression of many key regulatory and cellular signaling pathways in cancer cells *in vitro.*

Based on these studies (and others not presented here), it would appear that proanthocyanidins possess the ability to provide not only possible chemo-protective but also possible chemo-preventative effects / benefits against cancer development and progression and this supports the concept of inclusion of proanthocyanidin containing natural products and foods as a part of a healthy “anti-cancer ” diet.

REFERENCES

[1] Aron, P.M., Kennedy, J.A. Flavan-3-ols: Nature, occurrence and biological activity. *Mol.. Nutr, Food Res.* 2008, 52: 79-104.

[2] Mongas, M., Quintanilla-Lopez, J., Gomez-Cordoves, C., Bartolome,B., Lebron-Aguilar, R. MALDI-TOF MS Analysis of plant proanthocyanidins. *J. Pharm. Biomed. Anal.* 2010, 51: 358-372.

[3] Neuwirt, H., Arias, M.C., Pufr, M., Hobisch, A., Culig, Z. Oligomeric proanthocyanidin complexes (OPC) exert anti-proliferative and pro-apoptotic effect on prostate cancer cells. *Prostate* 2008, 68: 15, 1647-1654.

[4] Vayalil, P. K., Mittal, A., Katiyar, SK. Proanthocyanidins from grape seeds inhibit expression of matrix metalloproteinases in human prostate carcinoma cells which associated with the inhibition of activation of MAPK and NF kappa B. *Carcinogenesis* 2004, 25: 6, 987-995.

[5] Chen, Q., Liu, ZF., Zheng, PS. Grape seed proanthocyanidins (GSPs) inhibit the growth of cervical cancer by inducing apoptosis mediated by the mitochondrial pathway. *PLoS One* 2014 9,9, :e107045.

[6] Mantena, SK., Baliga, MS., Katiyar, SK. Grape seed proanthocyanidins induce apoptosis and inhibit metastasis of highly metastatic breast carcinoma cells. *Carcinogenesis* 2006, 27:8, 1682-1691.

[7] Song, X., Siriwardhana, N. Rathore, K., Lin,D., Wang, HC. Grape seed proanthocyanidin suppression of breast cancer cell carcinogenesis induced by chronic exposure to combined, 4-(methynitrosamino-)-1-(3-pyridyl) -1-butanone and benzo[a]pyrene. *Mol. Carcinog.*2010, 49:5, 450-463.

[8] Engelbrecht, AM., Mattheyse, M., Ellis, B., Loos, B., Thomas, M., Smith, R., Peters, R., Smith, C., Myburgh, K.Proanthocyanidins from grape seeds inactivates the PI3-kinase /PKB pathway and induces apoptosis in a colon cancer cell line. *Cancer Lett.* 2007, 258: 1, 144-153.

[9] Huang, S., Yang, N. Liu.,Y., Gao, J., Huang, T., Hu, L., Zhao, J., Li, Y., Zhang, Z. Grape seed proanthocyanidins inhibit colon cancer induced angiogenesis through suppressing the expression of VEGF and Ang1. *Int. J. Mol.Med.* 2012, 30 : 6, 1410-1416.

[10] D.Angelo. L., Piazzi, G., Pacilli, A., Prossomariti,A., Fazio, C., Montanaro, L., Graziani, G., Fogliano, V., Munarini, A., Bianchi, F., Belluzzi, A., Bazzoli,F., Ricciardiello, L. A combination of eicosapentaenoic acid-free fatty acid, epigallocatechin-3-gallate and proanthocyanidins has a strong effect on mTOR signaling in colorectal

cancer cells, *Carcinogenesis* 2014, Aug. 14 doi: 10.1093/carcino/bgu173. [Epub].

[11] Dass, K., Ahmad, A., Azmi, A.S., Sarkar, S.H., Sarkar, F.H. Evolving role of uPA/uPAR system in human cancers. *Cancer Treatment Reviews* 2008, 34: 122-136.

[12] Uchino, R., Madhayastha, R., Madhayastha, H., Dhungana, S., Nakajima, Y., Omura, S., Maruyama, M. NFkappa B-dependent regulation of urokinase plasminogen activator by proanthocyanidin-rich grape seed extract: effect on invasion by prostate cancer cells. *Blood Coagul. Fibrinolysis* 2010, 21 6: 528-533.

[13] Prasad, R., Vaid, M., Katiyar, SK. Grape proanthocyanidin inhibit pancreatic cancer cell growth in vitro and in vivo through the induction of apoptosis and by targeting the PI3K/Akt pathway. *PLoS One* 2014, 7, 8,: e43064 doi: 101371.

[14] Cedo, L., Castell-Auvi, A., Pallares, V., Macia, A., Blay, M., Ardevol, A., Motilva, MJ., Pinent, M. Gallic acid is an active component for the anticarcinogenic action of grape seed procyanidins in pancreatic cancer cells. *Nutr. Cancer* 2014, 66: 88-96.

[15] Prasad, R., Katiyar, SK. Grape seed proanthocyanidins inhibit migration potential of pancreatic cancer cells by promoting mesenchymal –to-epithelial transition and targeting NF-κB. *Cancer Lett.* 2012, 334: 1, 118-126.

[16] Punthil. T., Katiyar, SK. Inhibition of non-small lung cancer cell migration by grape seed proanthocyanidins is mediated through the inhibition of nitric oxide, guanylate cyclase and ERK 1/2. *Mol. Carcinog.* 2009, 48: 3, 232-242.

[17] Sharma, SD., Meeran, SM., Katiyar, SK. Proanthocyanidins inhibit in vitro and in vivo growth of human non-small cell lung cancer cells by inhibiting the prostaglandin E(2) and prostaglandin E(2) receptors. *Mol. Cancer Ther.* 2010, 9:3, 569-580.

[18] Vaid, M., Katiyar, SK. Grape seed proanthocyanidins inhibit cigarette smoke condensate induced lung cancer cell migration through inhibition of NADPH oxidase and reduction in the binding of p22phox and p47phox proteins. *Mol. Carcinog.* 2014, May 5, doi: 10.1002/mc.22173.

[19] Singh, T., Sharma, SD., Katiyar, SK. Grape proanthocyanidins induce apoptosis by loss of membrane potential of human non-small cell lung cancer cells in vitro and in vivo. *PLoS One* 2011, 6,11: e27444. doi : 10.1371.

[20] Vaid, M., Prasad, R., Singh, T., Jones, V., Katiyar, SK. Grape seed proanthocyanidins reactivate silenced tumor suppressor genes in human skin cancer cells by targeting epigenetic regulators. *Toxicol. Appl. Pharmacol.* 2012, 263:1, 122-130.

[21] Neto, C. Cranberry and blueberry: Evidence for protective effects against cancer and vascular diseases. *Mol. Nutr. Food Res.* 2007, 51: 652-664.

[22] Kandil, FE., Smith, MAL., Rogers, RB., Pepin, M-F., Song, LL., Pezzuto, JM., Seigler, DS. Composition of a chemopreventative proantho-cyanidin-rich fraction from cranberry fruits responsible for the inhibition of TPA-induced ODC activity. *J. Agric. Food Chem.* 2002, 50: 106301069.

[23] Pegg, AE. Regulation of ornithine decarboxylase. *J. Biol.Chem.* 2006, 281:14529-14532.

[24] Kresty, LA., Howell, AB., Baird, M. Cranberry proanthocyanidins induce apoptosis and inhibit acid-induced proliferation of human esophageal adenocarcinoma cells. *J. Agric. Food Chem.* 2008, 56: 3, 676-680.

[25] Liberty, AM., Amoroso, JW., Hart, PE., Neto, CC. Cranberry PACs and triterpenoids: anticancer activities in colon tumor cell lines. *Acta Hortic* 2009, 841: 61-66.

[26] Kresty, LA., Howell, AB., Baird, M. Cranberry proanthocyanidins mediate growth arrest of lung cancer cells through modulation of gene expression and rapid induction of apoptosis. *Molecules* 2011, 16: 2375-2390.

[27] Neto, CC., Krueger, CG., Lamoureaux, TL., Kondo, M. Vaisberg, AJ., Hurta, RAR., Curtis, S., Matchett, MD., Yeung, H., Sweeney, MI., Reed, JD. MALDI-TOF MS characterization of proanthocyanidins from cranberry fruit (*Vaccinium macrocarpon*) that inhibit tumor cell growth and matrix metalloproteinase expression *in vitro*. *J. Sci. Food Agric.* 2006, 86: 18-25.

[28] Stetler-Stevenson, WG., Yu, AE. Proteases in invasion: matrix metalloproteinases. Semin. *Cancer Biol.*2001, 11: 143-152.

[29] Deziel, BA., Patel, K., Neto, C., Gottschall-Pass, K., Hurta, RA. Proanthocyanidins from the American cranberry (*Vaccinium macrocarpon*) inhibit matrix metalloproteinase-2 and matrix metalloproteinase-9 activity in human prostate cancer cells via alterations in multiple cellular signalling pathways. *J. Cell. Biochem.* 2011, 111: 3, 742-754.

[30] Matchett, MD., MacKinnon, SL., Sweeney, MI., Gottschall-Pass, KT., Hurta, RAR. Blueberry flavonoids inhibit matrix metalloproteinase activity in DU145 human prostate cancer cells. *Biochem. Cell.Biol.* 2005, 83: 637-643.
[31] Matchett, MD., MacKinnon, SL., Sweeney, MI., Gottschall-Pass, KT., Hurta, RAR. Inhibition of matrix metalloproteinase activity in DU145 human prostate cancer cells by flavonoids from lowbush blueberry (*Vaccinium angustifolium*) : possible roles for protein kinase C and mitogen activated protein kinase mediated events. *J. Nutr. Biochem.* 2006, 17: 117-125.
[32] Kraft, TEB., Schmidt, BM., Knight, Y. Cuendet, M., Kang, Y-H., Pezzuto, JM., Seigler, DS., Lila, MA. Chemopreventative potential of wild lowbush blueberry fruits in multiple stages of carcinogenesis. *J Food Sci.* 2005, 70: S159-S166.
[33] Schmidt, BM., Erdman, JW., Lila, MA. Differential effects of blueberry proanthocyanidins on androgen sensitive and insensitive human prostate cancer cell lines, *Cancer Letters* 2006, 231: 240-246.
[34] Kim, J., McKeown, B., Jahic, H., Patel, K. Catalli, A., Kulka, M., Neto, C. Hurta, R. Proanthocyanidins from the American cranberry (*Vaccinium macrocarpon*) induce cell cycle alterations in DU145 human prostate cancer cells in vitro by affecting the expression of cell cycle associated proteins. *J. Functional Foods in Health and Disease* 2014, 4:2, 130-145.
[35] Scott, BE. Compounds of *Vaccinium macrocarpon* inhibit DU145 human prostate cancer cells. MSc. Thesis 2010, University of Prince Edward Island, Prince Edward Island, Canada.
[36] Kresty, LA., Clarke, J., Ezell, K., Exum, A., Howell, AB., Guettouche, T. MicroRNA alterations in Barrett's esophagus, esophageal adenocarcinoma and esophageal adenocarcinoma cell lines following cranberry extract treatment: insight for chemopreventation. *J. Carcinog.* 2011 10:34 doi: 10.4103/1477-3163.91110.
[37] Kim, KK. , Singh, AP., Singh, RK. Demartino, A., Brard, L., Vorsa, N., Lange, TS., Moore, RG. Anti-angiogenic activity of cranberry proanthocyanidins and cytotoxic properties in ovarian cancer cells. *Int. J. Oncol.* 2012, 40:1, 227-235.
[38] Singh, AP. Singh, RK. , Kim, KK., Satyan, KS., Nussbaum, R., Torres, M., Brard, L., Vorsa, N. Cranberry proanthocyanidins are cytotoxic to human cancer cells and sensitize platinum –resistant ovarian cancer cells to paraplatin. *Phytother. Res.* 2009, 23: 8, 1066-1074.

[39] Miura, T., Chiba, M., Kasai, K., Nozaka, H., Nakamura, T., Shoji, T., Kanda, T., Ohtake, Y., Saito, T. Apple procyanidins induced tumor cell apoptosis through mitochondrial pathway activation of caspase-3. *Carcinogenesis* 2008, 29:3, 585-593.

[40] Hibasami, H., Shoji, T., Shibuya, I., Higo, K., Kanda, T. Induction of apoptosis by three types of procyanidin isolated from apple (Rosaceae Malus pumila) in human stomach cancer KATO III cells. *Int. J. Mol. Med.* 2004, 13: 795-799.

[41] Gosse, F., Roussi, S., Guyot, S., Schoenfelder, A., Mann. A., Bergerat, JP., Seiler, N., Raul, F. Potentiation of apple procyanidin-triggered apoptosis by the polyamine oxidase activator MDL72527 in human colon cancer-derived metastatic cells. *Int. J Oncol.* 2006, 29: 2, 423-428.

[42] Gosse, F., Guyot, S., Roussi, S., Lobstein, A., Fischer, B., Seiler, N., Raul, F. Chemopreventative properties of apple procyanidins on human colon cancer-derived metastatic SW620 cells and in a rat model of colon carcinogenesis, *Carcinogenesis* 2005, 26: 7, 1291-1295.

[43] Teller, N., Roth, M., Esselen, M., Fridrich , D., Boettler, U., Blust, V., Will, F., Dietrich, H., Raul, F., Hummer, W., Richling, E., Schreier, P., Marko, D. Apple procyandins affect several members of the ErbB receptor tyrosine kinase family in vitro. *Food Funct.* 2013, 4: 5, 689-697.

In: Proanthocyanidins
Editor: Ingrid Sullivan
ISBN: 978-1-63463-430-4

Chapter 6

ANTI-AGING EFFECTS OF APPLE PROCYANIDINS

Tadahiro Sunagawa[1,2], Isao Masuda[3,4] and Takahiko Shimizu[2,4*]

[1]Research Laboratories for Alcohol Beverages, Asahi Breweries, Ltd., Ibaraki, Japan
[2]Molecular Gerontology, Tokyo Metropolitan Institute of Gerontology, Tokyo, Japan
[3]Research & Development Laboratories for Innovation, Asahi Group Holdings, Ltd., Ibaraki, Japan
[4]Department of Advanced Aging Medicine, Chiba University Graduate School of Medicine, Chiba, Japan

ABSTRACT

Apple polyphenols (APs) mainly consist of procyanidins (PCs), which are composed of polymerized (–)-epicatechins and (+)-catechins. APs are contained in the edible part of the apple, so apples may be good source of dietary polyphenols. APs have been reported to display several beneficial health effects. In this chapter, we summarize the anti-aging effects of APs on age-related diseases in model organisms. First,

* Correspondence: Takahiko Shimizu, Ph.D.; Department of Advanced Aging Medicine, Chiba University Graduate School of Medicine, 1-8-1 Inohana, Chuo-ku, Chiba 260-8670, Japan Telephone: +81-43-222-7171, Fax: +81-43-226-2095, E-mail: shimizut@chiba-u.jp

treatment with APs or PCs purified from APs extended the lifespan of *Caenorhabditis elegans* to a similar extent as treatment with *trans*-resveratrol. In contrast, treatment with APs had no effect on the longevity of *sir-2.1* worms, which lack the activity of SIR2, a member of the sirtuin family, indicating that APs showed SIR2-dependent anti-aging effects. In general, polyphenols have anti-oxidant activity. APs and PC dimers showed a higher level of superoxide dismutase-like activity and oxygen radical absorbance capacity than other polyphenols or antioxidants, as well as anti-oxidant effects on mouse myoblast cells *in vitro*. Next, we orally administered APs to heart/muscle-specific manganese-superoxide dismutase (Mn-SOD)-deficient mice, which exhibit lethal dilated cardiomyopathy due to intrinsic enhanced mitochondrial superoxide. Dietary APs significantly suppressed the progression of cardiac fibrosis, which extended the survival of the mutant mice. Dietary APs also suppressed oxidative DNA damage and reactive oxygen species production in enzymatically-dissociated cardiomyocytes of the mutant mice. Notably, dietary APs also improved the electrophysiological abnormalities, such as action potential and inwardly rectifier potassium current changes in the cardiomyocytes of the mutant mice, as determined using patch clamp techniques, and decreased the incidence of ventricular tachycardia or the induction of fibrillation using a burst ventricular pacing protocol. Furthermore, APs remarkably suppressed amyloid-β protein 42 (Aβ42) aggregation and dissociated Aβ42 aggregates *in vitro*, indicating that APs are potent suppressors of Aβ aggregation. APs also significantly inhibited Aβ neurotoxicity and stimulated the proliferation of neural PC-12 cells. Interestingly, treatment with APs also reduced polyglutamine aggregation in the *C. elegans* Q40 strain, a model of Huntington's disease. These findings strongly indicate that APs showed anti-aging effects, such as anti-oxidative, anti-protein aggregation and pro-longevity effects, using model organisms. Apple procyanidins are promising functional foods to prevent age-related disorders.

Introduction

Polyphenols include several groups of compounds, such as anthocyanins, flavonols and phenolic acids, which are accumulated in the plants themselves, as well as their fruits [1]. Several studies have reported that dietary polyphenols from plants prevent oxidative stress *in vivo* [2, 3]. Apples are one of the most common fruits grown and consumed worldwide, in addition to grapes, oranges and bananas [4]. The polyphenols extracted from apples (*Malus pumila* Mill., Rosaceae) mainly contain procyanidins (PCs), also called condensed tannins. PCs comprise approximately 65% of polyphenols from unripe apples, and the remaining approximately 35% of compounds are regarded as the monomer fraction (MNs) (Figure 1A) [5]. PCs are formed by catechin oligomers composed of (−)-epicatechin and (+)-catechin monomers (Figure 1B) [6]. Procyanidin B1 (PB1) is an epicatechin-catechin dimer, procyanidin B2 (PB2) is an epicatechin-epicatechin dimer, and procyanidin C1 (PC1) is an epicatechin-epicatechin-epicatechin trimer. PCs are a kind of combination of high-molecular-mass polyphenols, the molecular mass distribution of which can be profiled using a size-exclusion HPLC method [7]. The profile of apple polyphenols (APs) is different from that of green tea (*Camellia sinensis*) polyphenols, which consist of monomer catechins, mainly epigallocatechin gallate (EGCG) (Figure 1C).

PCs are also contained in other fruits, such as grape (*Vitis* spp.) seeds [8], which are included in red wine, with the relationship between the health benefits and the PCs content of wine having been reported previously [9]. However, the profile of APs is different from that of grape seed polyphenols (Figure 1C), with apple PCs consisting of a larger amount of low-degree polymerized oligomers and a smaller amount of high-degree polymerized oligomers.

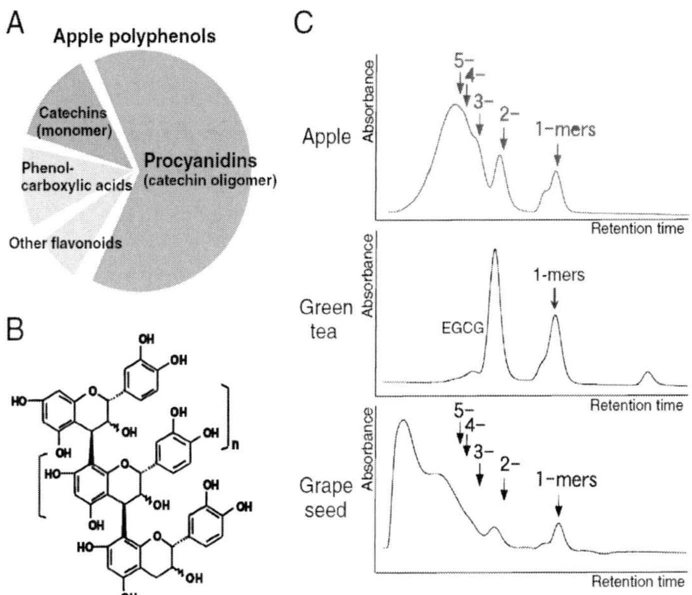

Figure 1. The composition of apple polyphenols (APs) (A), the chemical structure of apple procyanidins (PCs) (B) and the elution profiles of APs, tea polyphenols and grape seed polyphenols using gel chromatography (C) [7].

A structural analysis using NMR spectroscopy revealed that tetrameric PCs from unripe apples could be isolated, and the structures have been elucidated by NMR spectroscopy [10]. In addition, profiling of PC oligomer compounds from different apple varieties has been performing using comprehensive two-dimensional liquid chromatography [11]. The APs have been reported to exert various physiological functions, such as anti-tumor, anti-allergy and anti-obesity effects [12-14], as well as inhibitory effects on triglyceride adsorption, depending on the extent of polymerization of PCs in the human study [15]. However, the *in vivo* anti-aging effects, including the effects of the lifespan, redox balance and proteostasis, of APs and apple PCs have not been studied so far. In this chapter, we summarize the anti-aging effects of APs using model organisms for age-related diseases.

Apple Polyphenols Extend the Lifespan of *Caenorhabditis Elegans*

Caenorhabditis elegans is a simple multicellular organism with a short lifespan and generation time. Thus, *C. elegans* is widely used as an experimental model organism for studying aging and longevity. Dietary restriction extends the lifespan in *C. elegans,* as well as mammals [16]. We first evaluated the effects of APs and PCs purified from APs on the lifespan of *C. elegans*. Treatment with APs or PCs could significantly extend the mean lifespan of wild-type worms (Figure 2), to a similar extent as a *trans*-resveratrol [17]. In contrast, treatment groups given MNs purified from APs or epicatechin (the major component of the monomer fraction) did not show an extended lifespan [17], indicating that epicatechin oligomers, PCs, mainly contribute to the longevity effects of APs in worms. Wilson et al. have reported that blueberry extracts, which mainly contained PCs, also extended the lifespan of nematodes [18]. We thus concluded that PCs are a food factor regulating the lifespan of *C. elegans*.

It has been reported that *trans*-resveratrol extends the lifespan of several organisms in a SIR2-dependent manner [19, 20]. In order to investigate the involvement of SIR2 activity on the longevity effect of PCs in *C. elegans*, we examined the lifespan of *sir-2.1*worms, which are mutant worms that lack histone deacetylase (HDAC) SIR2 activity [19]. PCs could not extend the longevity of *sir-2.1* mutant worms, nor could *trans*-resveratrol. These findings indicated that PCs prolong the lifespan of worms in a SIR2-dependent manner. However, in an *in vitro* HDAC assay, PCs failed to activate SIRT1, in contrast with *trans*-resveratrol, showing that the mechanism underlying the longevity effect is not due to a SIRT1 agonist effect. Since the upregulation of the *sir-2* gene extends the survival in worms [21], PCs might regulate *sir-2* expression.

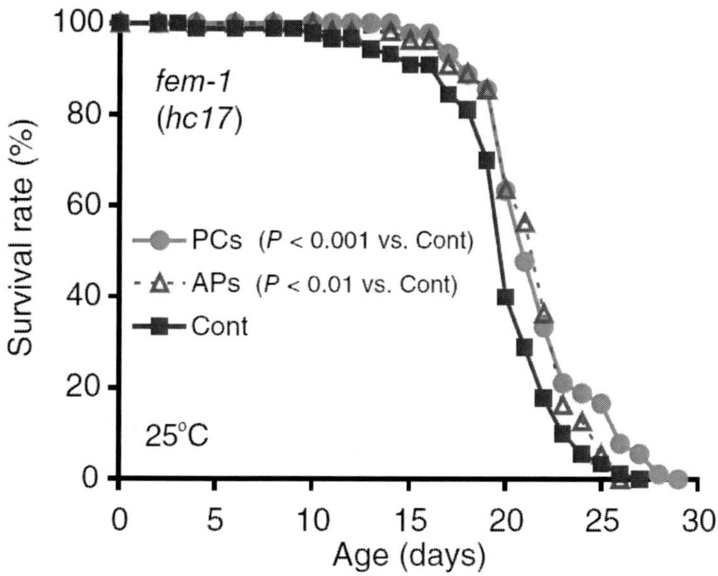

Figure 2. Apple PCs extend the lifespan of *Caenorhabditis elegans* [17]. The survival curves of *fem-1* worms treated with 65 μg/mL PCs ($n = 90$), 100 μg/mL APs ($n = 55$) and the control group ($n = 89$).

Apple Polyphenols Regulate Mitochondrial Superoxide Generation and Extend the Lifespan of a Mouse Model of Dilated Cardiomyopathy

A disturbance in the balance between the production of reactive oxygen species (ROS) and the antioxidant defense system causes oxidative stress, which leads to tissue injury and dysfunction [22,23]. It has been proposed that the accumulation of oxidative damage accelerates aging [24]. Moreover, in the heart, the generation of ROS and the age-related loss of antioxidant capacity might be involved in age-related cardiac dysfunction [25-27].

In order to examine the anti-oxidative effects of APs *in vitro*, we measured the superoxide dismutase (SOD)-like activity [28] and oxygen radical absorbance capacity (ORAC) [29] of APs and PCs. The SOD-like activity of APs, PC dimers, PB1 and PB2, and PC trimer, PC1, was higher than that of EUK-134, a mimetic of SOD (Figure 3A). In particular, PB1 and PB2 showed a high level of SOD-like activity. APs, PB1, PB2, and PC1, as well as *trans*-resveratrol, also showed a high level of ORAC compared to the antioxidant vitamin C or EGCG (Figure 3B).

In order to examine the anti-oxidative effects of APs *in vitro*, we then measured the intracellular ROS levels of C2C12 cells that had been treated with H_2O_2 using the fluorescent probe.

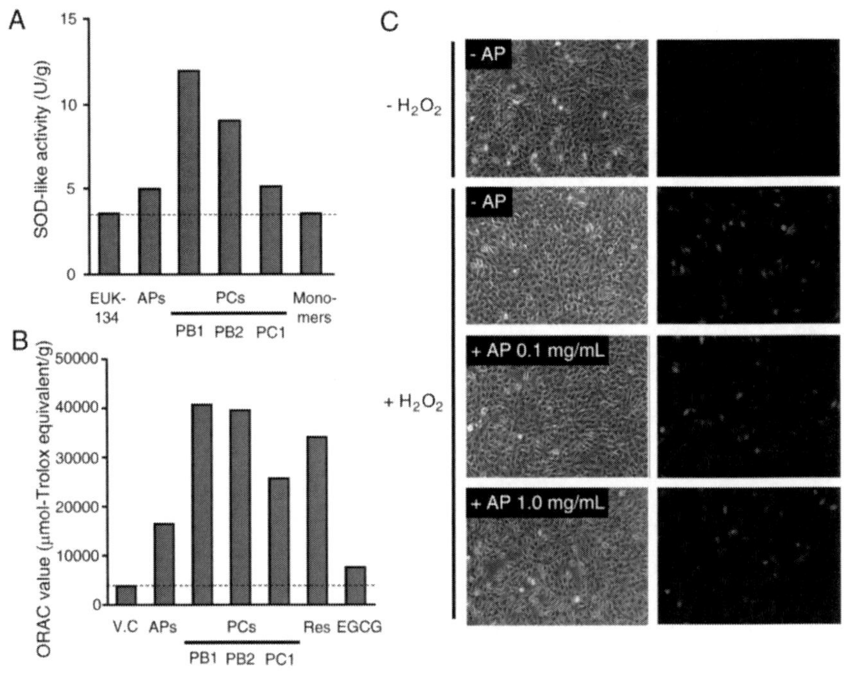

Figure 3. APs and apple PCs have anti-oxidant activity. The superoxide dismutase (SOD)-like activity of polyphenols (A), and the oxygen radical absorbance capacity (ORAC) values of polyphenols (B). The SOD-like activity was determined using the cellular xanthine/xanthine oxidase system as a superoxide source [28]. The ORAC assay was assessed according to the method described by Ishimoto et al. [29]. The 2,2'-azobis(2-amidinopropane)dihydrochloride was used as a peroxyl radical generator, and 6-hydroxy-2,3,7,8-tetramethylchroman-2-carboxylic acids (Trolox) was used as a standard. Monomers: monomer fraction of APs, V.C: vitamin C, Res: *trans*-resveratrol, EGCG: epigallocatechin gallate. The anti-oxidative effects of APs on mouse C2C12 myoblast cells (C) [30]. The cells were preincubated with APs and then were incubated with H_2O_2 and then with CM-H_2DCFDA. ROS production was induced in the cells treated with H_2O_2, as assessed by DCF fluorescence.

As a result, we found that preincubation with APs suppressed the production of ROS in C2C12 cells treated with H_2O_2 in a dose-dependent manner (Figure 3C). Next, in order to examine the anti-oxidative effects of APs against intrinsic ROS, we measured the intracellular ROS level of tamoxifen-induced manganese-superoxide dismutase (Mn-SOD) deficient mouse embryonic fibroblast (MEF) cells using the fluorescent probe. These studies showed that preincubation with APs suppressed the production of ROS in Mn-SOD-deficient MEFs [30].

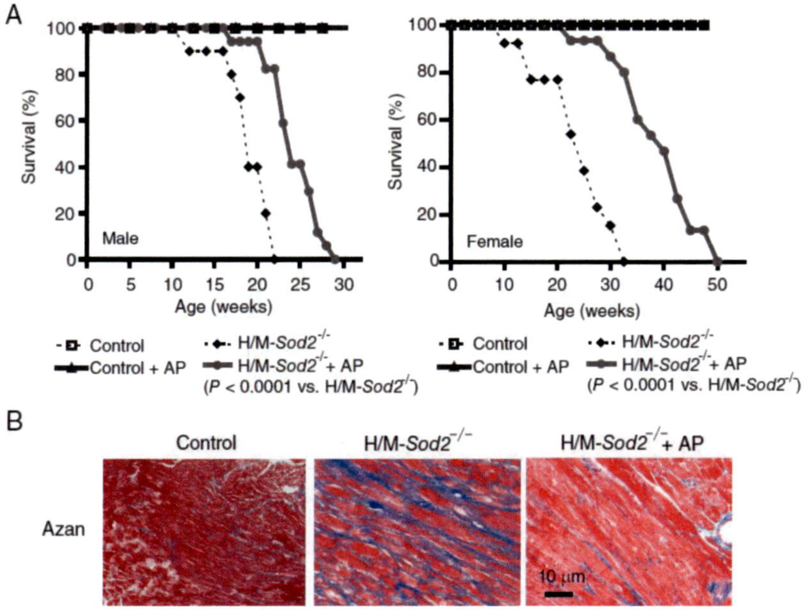

Figure 4. APs improve the survival and myocardial degeneration of H/M-*Sod2*$^{-/-}$ mice [30]. The survival curves of male and female H/M-*Sod2*$^{-/-}$ mice (A). The intake of APs extended the mean lifespan of the mice by 29% and 72% compared with water alone. Male H/M-*Sod2*$^{-/-}$; $n = 10$, H/M-*Sod2*$^{-/-}$+AP; $n = 17$, female H/M-*Sod2*$^{-/-}$; $n = 13$, H/M-*Sod2*$^{-/-}$+AP; $n = 15$. Azan-stained transverse sections of the left ventricular walls (B). The LV walls from male H/M-*Sod2*$^{-/-}$ mice showed myocardial degeneration, cardiomyocyte disarray and diffuse myocardial fibrosis. The intake of APs ameliorated the symptoms of the mutant mice.

Mn-SOD is an antioxidant enzyme that is located in mitochondria and catalyzes the dismutation of superoxide into oxygen and H_2O_2. We previously reported that conditional knockout mice that specifically lacked Mn-SOD in their cardiac and skeletal muscle developed progressive dilated cardiomyopathy (DCM) [31]. When APs were administered to heart and muscle Mn-SOD-deficient (H/M-*Sod2*$^{-/-}$) mice in drinking water containing 0.1% (w/v) APs, the APs significantly extended the lifespan of the male and female mutant mice (Figure 4A).

Dietary APs also significantly delayed their progressive body weight loss and cardiac dilatation. Moreover, APs treatment mitigated the accumulation of plasma creatine phosphokinase (CPK), the histopathology and the fibrosis in the heart of the mutant mice (Figure 4B). The biochemical analyses performed using ROS-reacting fluorescent reagents revealed that APs supplementation reduced the ROS production in the cardiomyocytes of mutant hearts *in vivo* [30]. Furthermore, dietary APs tended to decrease the 8-oxodG level of the nuclear fractions of the hearts of the H/M-*Sod2*$^{-/-}$ mice [30]. It is noteworthy that the concentration of APs in drinking water was 0.1% (w/v), which is equivalent to the level present in apple juice [32]. We previously reported that the administration of an SOD/catalase mimetic, EUK-8, significantly ameliorated the cardiac pathology and dysfunction of H/M-*Sod2*$^{-/-}$ mice by suppressing mitochondrial oxidative stress [33]. Taken together with our results, these data indicate that APs suppress mitochondrial superoxide production via their anti-oxidative effects, and prevent oxidative stress.

Apple Polyphenols Prevent Cardiac Electrophysiological Abnormalities of a High-Frequency Mouse Model of Ventricular Arrhythmias

Cardiac arrhythmia is a major health problem in the elderly. Although the incidence of ventricular arrhythmias is higher in patients with heart disease, premature ventricular contractions (PVCs) and multiform PVCs are common even in healthy aged people [34-36]. Oxidative stress participates in the age-related changes in the electromechanical function of the heart [25, 37, 38]. The surface electrocardiogram (ECG) showed that there was a significant prolongation in the QRS, PR and QT intervals and a flattening of the J wave observed in H/M-*Sod2*$^{-/-}$ mice [39]. The prolonged QT interval might stem from delayed repolarization of the ventricular action potentials and the increases in the duration of the QRS complex interval might reflect slowed conduction in mice. Dietary APs partially reversed the changes in the QRS interval in the ECG parameters of H/M-*Sod2*$^{-/-}$ mice, although it did not affect the ECG parameters of control mice [39].

The effective refractive period (ERP) from the left ventricle was significantly prolonged in the hearts of H/M-*Sod2*$^{-/-}$ mice relative to that of control mice [39], indicating that chronic exposure to oxidative stress affects the repolarization phase of the cardiac action potential. To determine whether chronic exposure to oxidative stress increases the susceptibility to ventricular arrhythmias, we also examined the induction of ventricular tachycardia (VT) or ventricular fibrillation (VF) using a burst ventricular pacing protocol (Figure 5A). In control mice, burst pacing failed to induce VT or VF, regardless of whether the mice received APs. In contrast, VT or VF could be induced in the H/M-*Sod2*$^{-/-}$ mice (Figure 5B). In APs-treated H/M-*Sod2*$^{-/-}$ mice, the ERP and incidence of VT or VF induction were significantly decreased, indicating that dietary APs improve the electrophysiological abnormalities arising due to chronic exposure to ROS.

Using whole-cell patch-clamp techniques with enzymatically-dissociated myocytes (Figure 5C), the membrane capacitance (Cm), which was used as an indirect index of the cell size, for the left ventricular myocytes of H/M-*Sod2*$^{-/-}$ mice was significantly greater than that of control mice, and dietary APs significantly reduced the Cm value of H/M-*Sod2*$^{-/-}$ myocytes [39]. The action potential durations (APD) at a 50% or 90% repolarization level in the left ventricular myocytes of H/M-*Sod2*$^{-/-}$ mice were significantly longer than those of control myocytes (Figure 5D). Dietary APs significantly reversed the prolongation of the APD in the left ventricular myocytes of H/M-*Sod2*$^{-/-}$ mice, although it did not influence the APDs in the control mice [39]. Abnormal automaticity, such as early afterdepolarizations, probably resulting from incomplete repolarization, was observed in about half of the ventricular cells of H/M-*Sod2*$^{-/-}$ mice, and dietary APs significantly inhibited the appearance of abnormal automaticity in H/M-*Sod2*$^{-/-}$ myocytes [39]. The dispersion of prolonged APDs, as well as the abnormal automaticity in the ventricular myocardium, might contribute to the increased susceptibility of H/M-*Sod2*$^{-/-}$ hearts to VT or VF.

There were no significant differences in the densities of whole-cell membrane currents, such as L-type Ca^{2+} or outward K^+ in the left ventricular myocytes isolated from H/M-*Sod2*$^{-/-}$ mice. In contrast, the density of the inwardly rectifier K^+ current (I_{K1}) was significantly lower in the left ventricular myocytes of H/M-*Sod2*$^{-/-}$ mice than in control mice, and dietary APs significantly increased the outward component of the I_{K1} at -50 mV in H/M-*Sod2*$^{-/-}$ ventricular cells, although it did not increase the density of I_{K1} in the ventricular cells of control mice (Figure 5E).

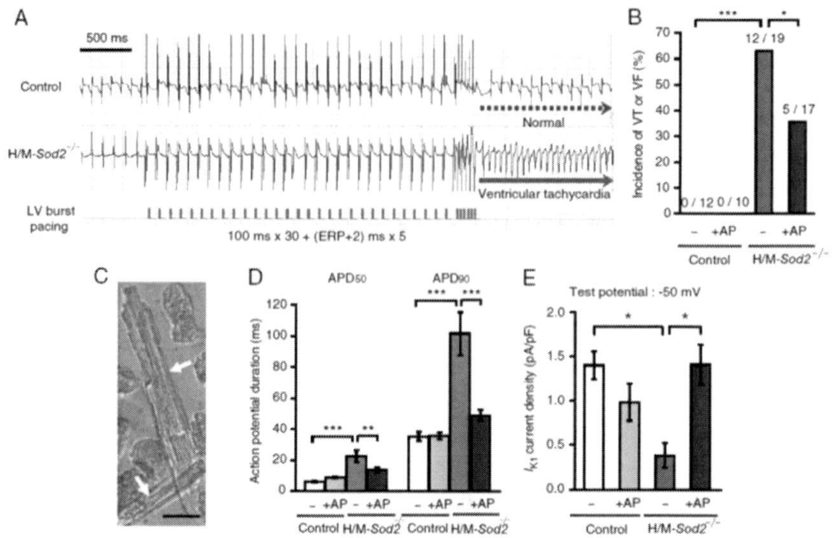

Figure 5. APs improve the cardiac electrophysiological abnormalities of H/M-*Sod2*$^{-/-}$ mice [39]. Representative ECG records during burst ventricular pacing observed in a control mouse and a H/M-*Sod2*$^{-/-}$ mouse (A). The incidence of ventricular tachycardia (VT) or ventricular fibrillation (VF) induced by burst pacing in control and H/M-*Sod2*$^{-/-}$ mice with and without AP treatment (B). Dietary APs decreased the incidence of VT or VF induction in H/M-*Sod2*$^{-/-}$ mice. Differences between groups, ***$P < 0.001$, *$P < 0.05$. The microscopic observation of enzymatically-dissociated left ventricular cardiomyocytes from H/M-*Sod2*$^{-/-}$ mice (C). The arrows show cardiomyocytes. The bar shows 20 μm. The action potential durations at 50% and 90% repolarization levels (APD50 and APD90) in the myocytes in each group (D). Control, $n = 49$; Control+AP, $n = 32$; H/M-*Sod2*$^{-/-}$, $n = 32$; and H/M-*Sod2*$^{-/-}$+AP, $n = 41$. ***$P < 0.001$, **$P < 0.01$. The density of the I_{K1} in H/M-*Sod2*$^{-/-}$ ventricular cells (E). The values are the means ± SEM, Control, $n = 32$; Control + AP, $n = 31$; H/M-*Sod2*$^{-/-}$, $n = 32$; and H/M-*Sod2*$^{-/-}$ + AP, $n = 32$. Differences between groups, *$P < 0.05$.

The inward rectifier K^+ channel provides one of the major components of the repolarizing outward current, the I_{K1}, in the cardiac action potential [40, 41]. The reduction in the I_{K1} may be involved in the prolongation of APDs, as well as the appearance of abnormal automaticity. Several studies demonstrated a significant suppression of the inwardly rectified K^+ channel by acute exposure to oxidative stress [42-44]. In the hearts of the Mn-SOD deficient mice, excess ROS reduced the I_{K1}, and dietary APs might prevent cardiac electrophysiological abnormalities due to a recovery of the I_{K1} as a result of their anti-oxidative effects.

Apple Polyphenols Suppress Protein Aggregation

Alzheimer's disease (AD) is a typical age-related and progressive neurodegenerative disease with memory impairment in later life. AD is diagnosed by the presence of amyloid accumulation, which is observed as a deposition in the hippocampus and cerebral cortex, frequently referred to as "senile plaque" [45]. Amyloid β-protein 42 (Aβ42), which consists of 42 residues, is observed mainly in the core of senile plaques. The protein forms strong self-aggregates that are associated with neurotoxicity *in vitro* [46]. In an Aβ42-thioflavin-T (Th-T) binding assay, APs and PCs dramatically abrogated the Aβ aggregation [47]. PCs inhibited the Aβ aggregation in a dose-dependent manner at an IC50 of 4.8 μg/mL. Aggregates were decreased by high-dose APs or PCs. PCs had approximately two-fold the anti-Aβ aggregative ability compared to APs. In contrast, MNs showed limited effects on Aβ aggregation.

We next confirmed the inhibitory effects of PCs on amyloid aggregation using a centrifugation method, to exclude the possibility that exogenous compounds affected the Th-T fluorescence intensity [48]. We observed the typical aggregates of Aβ after incubation (Figure 6A; vehicle). Treatment with 100 μg/mL APs and 65 μg/mL PCs completely diminished the aggregates (Figure 6A; APs and PCs), while 35 μg/mL MNs treatment did not (data not shown), demonstrating that PCs and APs exhibited strong anti-amyloidogenic ability *in vitro*.

PCs also dissociated Aβ42 aggregates following a post-treatment with polyphenols [47]. Although APs and PCs are orange-brown in color, we observed a brown-colored pellet that formed following treatment of Aβ42 aggregates with PCs and APs, suggesting that PCs could bind to Aβ42 aggregates. On the other hand, treatments with either vitamin-C or the potent antioxidative agent, EUK-134, did not suppress the Aβ aggregation, thus suggesting that the anti-aggregative activity of PCs is likely independent of its antioxidant properties. The chemical structures of these compounds might affect the Aβ aggregation, rather than their antioxidative capacity.

Aβ42 plays a pivotal role in the pathogenesis AD because of its potent aggregative ability and neurotoxicity [49]. PCs also had neuroprotective effects against Aβ42-induced neurotoxicity of neural PC-12 cells, as determined using a MTT assay (Figure 6B) [50]. PCs have a more potent ability to promote Aβ disaggregation and neuroprotection than whole APs. In contrast, MNs showed limited depression of Aβ aggregation.

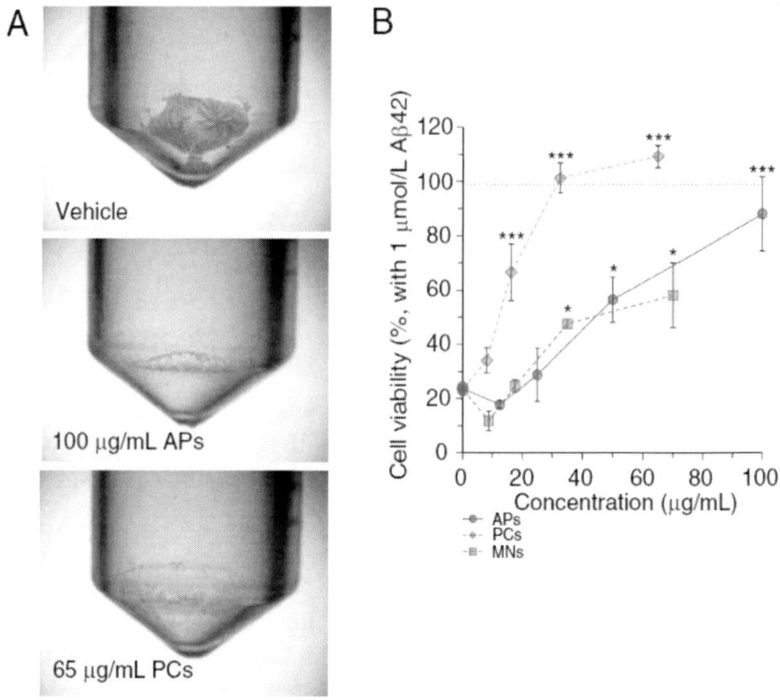

Figure 6. Apple PCs suppress and extinguish Aβ aggregation *in vitro* [47]. Microscopic observations of Aβ42 precipitates by centrifugation in the presence of polyphenols (A). The viability of PC-12 cells treated with Aβ42 (B). Cells were pretreated with various concentrations of APs, PCs and MNs, and then were incubated with Aβ42. PCs restored the viability of PC-12 cells treated with Aβ42. The values are the means ± SEM, $n = 3$. Compared to the control group, ***$P < 0.001$, and *$P < 0.05$.

It is possible that whole APs and MNs may contain factors that counteract the anti-aggregative and neuroprotective activity of PCs. Grape seed polyphenolic extracts including PCs interfere with paired helical filament formation by direct physical intercalation with taumolecules [51]. Therefore, apple PCs might have a similar suppressive effect on intermolecular aggregation of tau, as well as Aβ42.

Huntington's disease (HD) is also a progressive neurodegenerative disease, and is diagnosed by the presence of polyglutamine (polyQ) aggregation observed in addition to the atrophy of the cerebral cortex or basal ganglion [52]. Interestingly, PCs treatment also reduced polyQ deposition in body wall muscle cells of the transgenic *C. elegans* Q40 strain, a model of HD (Figure 7A, B) [53], indicating that apple PCs inhibited endogenous polyQ aggregation *in vivo*.

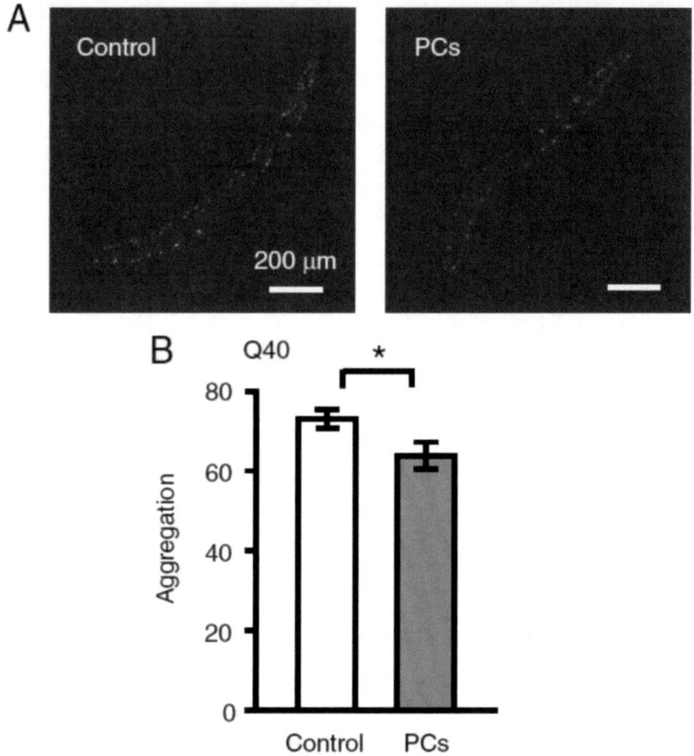

Figure 7. Apple PCs reduce the polyglutamine (polyQ) aggregation in *C. elegans*. The *C. elegans* Q40 strain was treated with PCs (A). Q40-YFP deposits were observed as dots. The number of aggregates of Q40 worms treated with PCs (12-day-old) (B). PCs reduced the aggregate formation in the worms. The values are the means ± SEM, control; $n = 18$, PCs; $n = 9$. Compared to the control group, $^{*}P < 0.05$.

Apple Polyphenols Are Good Food Factors to Prevent Age-Related Disorders

The current research shows that both nutritional apple- and pharmaceutical statin-based approaches may have the potential to reduce mortality due to vascular disease, recalling the phrase that "An apple a day keeps the doctor away" [54]. Recently, Caton et al. report that apple PCs showed beneficial effects on vascular endothelial function [55], as well as an inhibitory effect on lectin-like oxidized LDL receptor-1, which is a key player in the development of atherosclerosis [56]. It also showed activating effects on K^+ channels of aortic endothelial cells [57], and anti-inflammatory effects on intestinal epithelial cells [58]. The tetrameric PC increases the levels of a gastrointestinal hormone, GLP-1, and insulin secretion in mice, showing that PCs act as agonists in the gut [59]. Further studies are necessary to elucidate the influence of the degree of PCs polymerization on their favorable effects.

This chapter showed that APs and apple PCs have several anti-aging effects on organisms *in vitro* and *in vivo* (Figure 8). APs had a potent *in vivo* cardioprotective effect in a mouse model of DCM and ventricular arrhythmia, and extended the lifespan of the mice. APs mainly contain PCs, which have potent antioxidative activity [60]. The bioavailability of APs has been reported previously [61]. Shoji et al. revealed that apple PCs, which range in length from dimers to pentamers, are detectable in rat plasma after their oral administration. Since the ROS levels were lower in the ventricular cells of dietary APs 0.1% (w/v) water-treated H/M-*Sod2*$^{-/-}$ mice than in untreated H/M-*Sod2*$^{-/-}$ mice, the favorable cardioprotective effects might be ascribed to the antioxidant effects of APs.

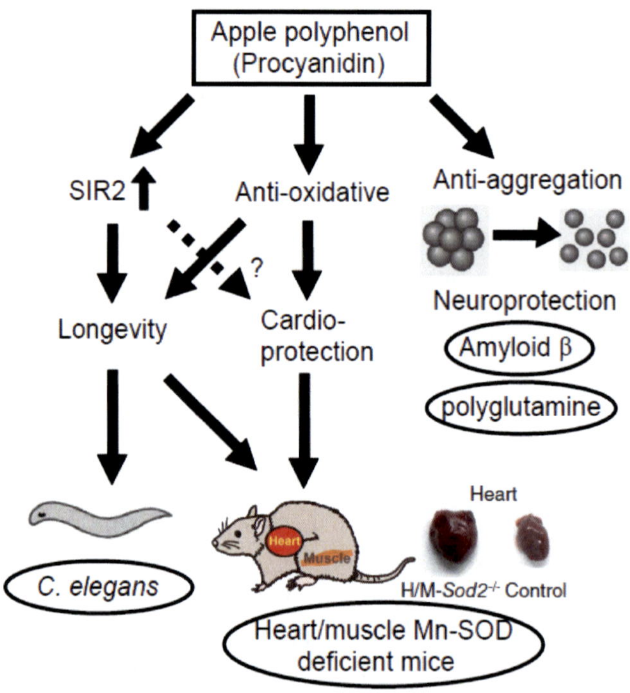

Figure 8. The APs and apple PCs have several anti-aging effects on organisms *in vitro* and *in vivo*.

In this chapter, we also revealed that APs showed SIR2-dependent anti-aging effects leading to lifespan extension in a nematode. SIRT1 may have an important role in the ability of the compounds to protect against cardiac ischemia [62]. It has been demonstrated that APs reduce the oxidative stress and inflammation involved in inflammatory bowel diseases, and these actions were accompanied by the induction of Nrf2 (orchestrating cellular antioxidant defenses and maintaining redox homeostasis) [63]. Taken together, these finding led us to propose that the beneficial effects of APs on lifespan involve not only a direct radical scavenging action, but also the modulation of activity of sirtuins, and/or an indirect anti-oxidative effect via the upregulation of cellular defense systems in cardiomyocytes. Further studies are necessary to elucidate the molecular mechanisms responsible for the effects of APs.

Apples are one of the major fruits cultivated and consumed all over the world. APs are contained in the edible part of the apple, and so are easier to routinely ingest compared to other active polyphenols, such as *trans*-resveratrol and grape seed polyphenols [64, 65]. Apple polyphenols and apple procyanidins may be promising and potential functional food factors to prevent or treat age-related disorders.

CONCLUSION

This chapter shows that apple polyphenols mainly consist of procyanidins have the *in vitro* and *in vivo* anti-aging effects, including the effects of the lifespan, redox balance and proteostasis. We propose that the beneficial effects on the lifespan involve not only a direct radical scavenging action, but also the modulation of activity of sirtuins, and/or an indirect anti-oxidative effect. Apple polyphenols may be promising and potential functional food factors to prevent or treat age-related disorders.

REFERENCES

[1] Tsao, R. (2010). Chemistry and biochemistry of dietary polyphenols. *Nutrients,* 2, 1231-1246.

[2] Pietta, P. G. (2000). Flavonoids as antioxidants. *J. Nat. Prod.*, 63, 1035-1042.

[3] Mursu, J., Voutilainen, S., Nurmi, T., Tuomainen, T. P., Kurl, S. & Salonen, J. T. (2008). Flavonoid intake and the risk of ischaemic stroke and CVD mortality in middle-aged Finnish men: the Kuopio Ischaemic Heart Disease Risk Factor Study. *Br. J. Nutr.*, 100, 890-895.

[4] Food and Agriculture Organization of the United Nations, FAOSTAT database (FAOSTAT, 2012). Available at http://faostat.fao.org/.

[5] Shoji, T., Masumoto, S., Moriichi, N., Kanda, T. & Ohtake, Y. (2006). Apple (*Malus pumila*) procyanidins fractionated according to the degree of polymerization using normal-phase chromatography and characterized by HPLC-ESI/MS and MALDI-TOF/MS. *J. Chromatogr.* A, 1102, 206-213.

[6] Ohnishi-Kameyama, M., Yanagida, A., Kanda, T. & Nagata, T. (1997). Identification of catechin oligomers from apple (*Malus pumila* cv. Fuji) in matrix-assisted laser desorption/ionization time-of-flight mass spectrometry and fast-atom bombardment mass spectrometry. *Rapid Commun. Mass Spectrom.*, 11, 31-36.

[7] Yanagida, A., Shoji, T. & Kanda, T. (2002). Characterization of polymerized polyphenols by size-exclusion HPLC. *Biosci. Biotechnol. Biochem.*, 66, 1972-1975.

[8] Santos-Buelga, C. & Scalbert, A. (2000). Proanthocyanidins and tannin-like compounds: Nature, occurrence, dietary intake and effects on nutrition and health. *J. Sci. Food Agric.*, 80, 1094-1117.

[9] Corder, R., Mullen, W., Khan, N. Q., Marks, S. C., Wood, E. G., Carrier, M. J. & Crozier, A. (2006). Oenology: red wine procyanidins and vascular health. *Nature,* 444, 566.

[10] Nakashima, S., Oda, C., Masuda, S., Tagashira, M. & Kanda, T. (2012). Isolation and structure elucidation of tetrameric procyanidins from unripe apples (*Malus pumila* cv. Fuji) by NMR spectroscopy. *Phytochemistry,* 83, 144-152.

[11] Montero, L., Herrero, M., Ibáñez, E. & Cifuentes, A. (2013). Profiling of phenolic compounds from different apple varieties using comprehensive two-dimensional liquid chromatography. *J. Chromatogr. A*, 1313, 275-283.

[12] Miura, T., Chiba, M., Kasai, K., Nozaka, H., Nakamura, T., Shoji, T., Kanda, T., Ohtake, Y. & Sato, T. (2008). Apple procyanidins induce tumor cell apoptosis through mitochondrial pathway activation of caspase-3. *Carcinogenesis,* 29, 585-593.

[13] Akiyama, H., Sato, Y., Watanabe, T., Nagaoka, M. H., Yoshioka, Y., Shoji, T., Kanda, T., Yamada, K., Totsuka, M., Teshima, R., Sawada, J.,

Goda, Y. & Maitani, T. (2005). Dietary unripe apple polyphenol inhibits the development of food allergies in murine models. *FEBS Lett.,* 579, 4485-4491.

[14] Osada, K., Funayama, M. & Fuchi, S. (2006). Effects of dietary procyanidins and tea polyphenols on adipose tissue mass and fatty acid metabolism in rats on a high fat diet. *J. Oleo Sci.,* 55, 79-89.

[15] Sugiyama, H., Akazome, Y., Shoji, T., Yamaguchi, A., Yasue, M., Kanda, T. & Ohtake, Y. (2007). Oligomeric procyanidins in apple polyphenol are main active components for inhibition of pancreatic lipase and triglyceride absorption. *J. Agric. Food Chem.*, 55, 4604-4609.

[16] Lakowski, B. & Hekimi, S. (1998). The genetics of caloric restriction in *Caenorhabditis elegans. Proc. Natl. Acad. Sci. USA,* 95, 13091-13096.

[17] Sunagawa, T., Shimizu, T., Kanda, T., Tagashira, M., Sami, M. & Shirasawa, T. (2011). Procyanidins from apples (*Malus pumila* Mill.) extend the lifespan of *Caenorhabditis elegans. Planta Med.,* 77, 122-127.

[18] Wilson, M. A., Shukitt-Hale, B., Kalt, W., Ingram, D. K., Joseph, J. A. & Wolkow, C. A. (2006). Blueberry polyphenols increase lifespan and thermotolerance in *Caenorhabditis elegans. Aging Cell*, 5, 59-68.

[19] Wood, J. G., Rogina, B., Lavu, S., Howitz, K., Helfand, S. L., Tatar, M. & Sinclair, D. (2004). Sirtuin activators mimic caloric restriction and delay ageing in metazoans. *Nature*, 430, 686-689.

[20] Valenzano, D. R., Terzibasi, E., Genade, T., Cattaneo, A., Domenici, L. & Cellerino, A. (2006). Resveratrol prolongs lifespan and retards the onset of age-related markers in a short-lived vertebrate. *Curr. Biol.*, 16, 296-300.

[21] Tissenbaum, H. A. & Guarente, L. (2001). Increased dosage of a sir-2 gene extends lifespan in *Caenorhabditis elegans*. Nature, 410, 227-230.

[22] Dröge, W. (2002). Free radicals in the physiological control of cell function. *Physiol. Rev.,* 82, 47-95.

[23] Cooke, M. S., Evans, M. D., Dizdaroglu, M. & Lunec, J. (2003). Oxidative DNA damage: mechanisms, mutation, and disease. *FASEB J.,* 17, 1195-1214.

[24] Finkel, T. & Holbrook, N. J. (2000). Oxidants, oxidative stress and the biology of ageing. *Nature,* 408, 239-247.

[25] Lucas, D. T. & Szweda, L. I. (1998). Cardiac reperfusion injury: aging, lipid peroxidation, and mitochondrial dysfunction. *Proc. Natl. Acad. Sci. USA,* 95, 510-514.

[26] Dhalla, N. S., Temsah, R. M. & Netticadan, T. (2000). Role of oxidative stress in cardiovascular diseases. *J. Hypertens.,* 18, 655-673.

[27] Lakatta, E. G. (2003). Arterial and cardiac aging: major shareholders in cardiovascular disease enterprises: Part III: cellular and molecular clues to heart and arterial aging. *Circulation*, 107, 490-497.

[28] Peskin, A. V. & Winterbourn, C. C. (2000). A microtiter plate assay for superoxide dismutase using a water-soluble tetrazolium salt (WST-1). *Clin. Chim. Acta*, 293, 157-166.

[29] Ishimoto, H., Tai, A., Yoshimura, M., Amakura, Y., Yoshida, T., Hatano, T. & Ito, H. (2012). Antioxidative properties of functional polyphenols and their metabolites assessed by an ORAC assay. *Biosci. Biotechnol. Biochem.,* 76, 395-399.

[30] Sunagawa, T., Watanabe, K., Ozawa, Y., Nakashima, S., Kanda, T., Tagashira, M., Sami, M., Kaneko, T., Tahara, S., Nakaya, H., Shirasawa, T. & Shimizu, T. (2012). Apple polyphenols regulate mitochondrial superoxide generation and extend survival in a mouse model of dilated cardiomyopathy. *Int. J. Life Sci. Med. Res.*, 2, 46-51.

[31] Nojiri, H., Shimizu, T., Funakoshi, M., Yamaguchi, O., Zhou, H., Kawakami, S., Ohta, Y., Sami, M., Tachibana, T., Ishikawa, H., Kurosawa, H., Kahn, R. C., Otsu, K. & Shirasawa, T. (2006). Oxidative stress causes heart failure with impaired mitochondrial respiration. *J. Biol. Chem.*, 281, 33789-33801.

[32] Kahle, K., Kraus, M. & Richling, E. (2005). Polyphenol profiles of apple juices. *Mol. Nutr. Food Res.,* 49, 797-806.

[33] Kawakami, S., Matsuda, A., Sunagawa, T., Noda, Y., Kaneko, T., Tahara, S., Hiraumi, Y., Adachi, S., Matsui, H., Ando, K., Fujita, T., Maruyama, N., Shirasawa, T. & Shimizu, T. (2009). Antioxidant, EUK-8, prevents murine dilated cardiomyopathy. *Circ. J.,* 73, 2125-2134.

[34] Glasser, S. P., Clark, P. I. & Applebaum, H. J. (1979). Occurrence of frequent complex arrhythmias detected by ambulatory monitoring: findings in an apparently healthy asymptomatic elderly population. *Chest,* 75, 565-568.

[35] Aronow, W. S., Epstein, S., K.S., S. & Koenigsberg, M. (1987). Prevalence of arrhythmias detected by ambulatory electrocardiographic monitoring and of abnormal left ventricular ejection fraction in persons older than 62 years in a long-term health care facility. *Am. J. Cardiol.,* 59, 368-369.

[36] Tresch, D. D. & Thakur, R. K. (1998). Ventricular arrhythmias in the elderly. *Emerg. Med. Clin. North Am.,* 16, 627-648.

[37] Lakatta, E. G. (2003). Arterial and cardiac aging: Major shareholders in cardiovascular disease enterprises. Part III: Cellular and molecular clues to heart and arterial aging. *Circulation*, 107, 490-497.

[38] van der Loo, B., Bachschmid, M., Spitzer, V., Brey, L., Ullrich, V. & Luscher, T. F. (2003). Decreased plasma and tissue levels of vitamin C in a rat model of aging: Implications for antioxidative defense. *Biochem. Biophys. Res. Commun.,* 303, 483-487.

[39] Sunagawa, T., Shimizu, T., Matsumoto, A., Tagashira, M., Kanda, T., Shirasawa, T. & Nakaya, H. (2014). Cardiac electrophysiological alterations in heart/muscle-specific manganese-superoxide dismutase-deficient mice: Prevention by a dietary antioxidant polyphenol. *BioMed Res. Int.*, 2014, 704291.

[40] Brouillette, J., Clark, R. B., Giles, W. R. & Fiset, C. (2004). Functional properties of K^+ currents in adult mouse ventricular myocytes. *J. Physiol.,* 559, 777-798.

[41] Brunet, S., Aimond, F., Li, H., Guo, W., Eldstrom, J., Fedida, D., Yamada, K. A. & Nerbonne, J. M. (2004). Heterogeneous expression of repolarizing, voltage-gated K^+ currents in adult mouse ventricles. *J. Physiol.*, 559, 103-120.

[42] Matsuura, H. & Shattock, M. J. (1991). Effects of oxidant stress on steady-state background currents in isolated ventricular myocytes. *Am. J. Physiol.,* 261, H1358-H1365.

[43] Nakaya, H., Takeda, Y., Tohse, N. & Kanno, M. (1992). Mechanism of the membrane depolarization induced by oxidative stress in guinea-pig ventricular cells. *J. Mol. Cell Cardiol.*, 24, 523-534.

[44] Bhatnagar, A. (1995). Electrophysiological effects of 4-hydroxynonenal, an aldehydic product of lipid peroxidation, on isolated rat ventricular myocytes. *Circ. Res.,* 76, 293-304.

[45] Selkoe, D. J. (2007). Developing preventive therapies for chronic diseases: lessons learned from Alzheimer's disease. *Nutr. Rev., 65*, S239-S243.

[46] Murakami, K., Irie, K., Morimoto, A., Ohigashi, H., Shindo, M., Nagao, M., Shimizu, T. & Shirasawa, T. (2003). Neurotoxicity and physicochemical properties of Abeta mutant peptides from cerebral amyloid angiopathy: implication for the pathogenesis of cerebral amyloid angiopathy and Alzheimer's disease. *J. Biol. Chem.*, 278, 46179-46187.

[47] Toda, T., Sunagawa, T., Kanda, T., Tagashira, M., Shirasawa, T. & Shimizu, T. (2011). Apple procyanidins suppress amyloid β-protein aggregation. *Biochem. Res. Int.*, 2011, 784698.

[48] Hudson, S. A., Ecroyd, H., Kee, T. W. & Carver, J. A. (2009). The thioflavin T fluorescence assay for amyloid fibril detection can be biased by the presence of exogenous compounds. *FEBS J.,* 276, 5960-5972.

[49] Murakami, K., Masuda, Y., Shirasawa, T., Shimizu, T. & Irie, K. (2010). The turn formation at positions 22 and 23 in the 42-mer amyloid β peptide: The emerging role in the pathogenesis of Alzheimer's disease. *Geriatr Gerontol Int,* 10, S169-S179.

[50] Behl, C., Davis, J. B., Lesley, R. & Schubert, D. (1994). Hydrogen peroxide mediates amyloid β protein toxicity. *Cell*, 77, 817-827.

[51] Pasinetti, G. M., Ksiezak-Reding, H., Santa-Maria, I., Wang, J. & Ho, L. (2010). Development of a grape seed polyphenolic extract with anti-oligomeric activity as a novel treatment in progressive supranuclear palsy and other tauopathies. *J. Neurochem.*, 114, 1557-1568.

[52] MacDonald, M. E., Ambrose, C. M., Duyao, M. P., Myers, R. H., Lin, C., Srinidhi, L., Barnes, G., Taylor, S. A., James, M. & Groot, N., et al. (1993). A novel gene containing a trinucleotide repeat that is expanded and unstable on Huntington's disease chromosomes. *Cell*, 72, 971-983.

[53] Morley, J. F., Brignull, H. R., Weyers, J. J. & Morimoto, R. I. (2002). The threshold for polyglutamine-expansion protein aggregation and cellular toxicity is dynamic and influenced by aging in *Caenorhabditis elegans*. *Proc. Natl. Acad. Sci. USA*, 99, 10417-10422.

[54] Briggs, A. D. M., Mizdrak, A. & Scarborough, S. P. (2013). A statin a day keeps the doctor away: comparative proverb assessment modelling study. *BMJ*, 347, f7267.

[55] Caton, P. W., Pothecary, M. R., Lees, D. M., Khan, N. Q., Wood, E. G., Shoji, T., Kanda, T., Rull, G. & Corder, R. (2010). Regulation of vascular endothelial function by procyanidin-rich foods and beverages. *J. Agric. Food Chem.,* 58, 4008-4013.

[56] Nishizuka, T., Fujita, Y., Sato, Y., Nakano, A., Kakino, A. & Ohshima, S. (2011). Procyanidins are potent inhibitors of LOX-1: a new player in the French Paradox. *Proc. Jpn. Acad. Ser. B Phys. Biol. Sci.*, 87, 104-113.

[57] Byun, E. B., Ishikawa, T., Suyama, A., Kono, M., Nakashima, S., Kanda, T., Miyamoto, T. & Matsui, T. (2012). A procyanidin trimer, C1, promotes NO production in rat aortic endothelial cells via both

hyperpolarization and PI3K/Akt pathways. *Eur. J. Pharmacol.*, 692, 52-60.

[58] Yoshioka, Y., Akiyama, H., Nakano, M., Shoji, T., Kanda, T., Ohtake, Y., Takita, T., Matsuda, R. & Maitani, T. (2008). Orally administered apple procyanidins protect against experimental inflammatory bowel disease in mice. *Int. Immunopharmacol.*, 8, 1802-1807.

[59] Yamashita, Y., Okabe, M., Natsume, M. & Ashida, H. (2013). Cinnamtannin A2, a tetrameric procyanidin, increases GLP-1 and insulin secretion in mice. *Biosci. Biotechnol. Biochem.*, 77, 888-891.

[60] Williamson, G. & Manach, C. (2005). Bioavailability and bioefficacy of polyphenols in humans. II. Review of 93 intervention studies. *Am. J. Clin. Nutr.*, 81, 243S-255S.

[61] Shoji, T., Masumoto, S., Moriichi, N., Akiyama, H., Kanda, T., Ohtake, Y. & Goda, Y. (2006). Apple procyanidin oligomers absorption in rats after oral administration: analysis of procyanidins in plasma using the porter method and high-performance liquid chromatography/tandem mass spectrometry. *J. Agric. Food Chem.*, 54, 884-892.

[62] Shinmura, K., Tamaki, K. & Bolli, R. (2008). Impact of 6-mo caloric restriction on myocardial ischemic tolerance: possible involvement of nitric oxide-dependent increase in nuclear Sirt1. *Am. J. Physiol. Heart Circ. Physiol.*, 295, H2348-H2355.

[63] Denis, M. C., Furtos, A., Dudonné, S., Montoudis, A., Garofalo, C., Desjardins, Y., Delvin, E. & Levy, E. (2013). Apple peel polyphenols and their beneficial actions on oxidative stress and inflammation. *PLoS One,* 8, e53725.

[64] Lu, Y. & Yeap Foo, L. (2000). Antioxidant and radical scavenging activities of polyphenols from apple pomace. *Food Chem.*, 68, 81-85.

[65] Oszmianski, J., Wojdylo, A. & Kolniak, J. (2009). Effect of enzymatic mash treatment and storage on phenolic composition, antioxidant activity, and turbidity of cloudy apple juice. *J. Agric. Food Chem.*, 57, 7078-7085.

INDEX

C

D

E

F

G

H

I

J

K

L

M

N

O

P

Q

R

S

T

U

V